T0320551

Error Estimation in Reactor Shielding Calculations

Originally published as *Pogreshnosti raschetov zashchity ot izlucheniĭ.*

© Energoatomizdat, 1983

Library of Congress Cataloging in Publication Data

Pogreshnosti raschetov zashchity ot izlucheniĭ.
    English.
    Error estimation in reactor shielding calculations.

Translation of: Pogreshnosti raschetov zashchity ot izlucheniĭ.
    1. Shielding (Radiation)—Design and construction.
    I. Boliâtko, V. V. (Viktor Viktorovich) II. Mashkovich, V. P. (Vadim Pavlo-
vich) III. Title. IV. Title: Error estimation in reactor shielding calculations.
V. Series.
    TK9210.P6413   1987   621.48'323   87-19290
    ISBN 0-88318-536-9

Originally published as *Pogreshnosti raschetov zashchity ot izluchenii̯.*

© Energoatomizdat, 1983

Library of Congress Cataloging in Publication Data

Pogreshnosti raschetov zashchity ot izluchenii̯.
    English.
    Error estimation in reactor shielding calculations.

(American Institute of Physics translation series)
Translation of: Pogreshnosti raschetov zashchity ot izluchenii̯.
    1. Shielding (Radiation)—Design and construction.
    I. Boliâtko, V. V. (Viktor Viktorovich) II. Mashkovich, V. P. (Vadim Pavlo-
vich) III. Title. IV. Title: Error estimation in reactor shielding calculations.
V. Series.
    TK9210.P6413   1987   621.48′323   87-19290
    ISBN 0-88318-536-9

Contents

Chapter 1. Basic concept, definitions, and approaches 1

Chapter 2. Calculation methods and models

Chapter 3. Errors in the results of shielding calculations and sensitivity of the results to the input parameters

Chapter 4. Experiments in sensitivity analysis and in the determination of calculation errors

Preface

The errors and uncertainties in calculation of the radiation fields in problems of shielding against ionizing radiation are studied. The results presented here can be used to develop more economical installations through reduction in the cost of shielding. This book is the first to appear on the topic anywhere in the world. It can be used extensively as a handbook and guide. It is addressed to physicist-engineers, designers, and workers concerned with radiation transport and shielding in installations with radiation sources.

Introduction

The development of radiation shielding may be broken up somewhat arbitrarily into four historical stages.

The first, which lasted about 50 years, stretched from the discovery of x rays and radioactivity down to the development of nuclear reactors. This was a stage in which the relatively low power of the radiation sources made it a relatively simple matter to shield the body against radiation, so there was of course no stimulus for an active research program on shielding.

The advent of nuclear reactors opened up a new stage in the development of radiation shielding. Science and industry demanded quick answers to the new questions of radiation safety which were continually arising. The basic thrust in this stage was to develop simplified, semiempirical, and empirical models for shielding calculations. For example, calculations in the exponential approximation using experimental values of the relaxation length were carried out widely in these years. At the same time, rigorous methods for solving the radiation-transport equation began to be developed. Some specialists regarded simplified methods as adequate for solving practical problems; other specialists—the majority—asked, "When will we see the end of empiricism and semiempiricism?" This stage may be regarded as the period in which the physics of radiation shielding finally emerged as an independent field of applied nuclear physics.

The beginning of the third stage can be put at the time of the resolution of the dispute between adherents of empirical and semiempirical approaches and the adherents of rigorous methods. The dispute was resolved in the favor of the latter approach. In this period there was a rapid development of numerical methods for solving the multigroup transport equation and of the Monte Carlo method.

Inextricably linked with the confirmation of nuclear energetics as a promising new source of electric power, the emergence of nuclear energetics as a major field of power production, and the increasing use of ionizing-radiation sources in the economy are the tasks of further improving radiation safety and of planning and developing the best biological shields. These goals require sharply improving the efficiency and quality of the multifaceted effort to solve problems of shielding physics. At this point we cannot be content to simply obtain a result. We must find the optimal solution of practical problems, including the choice of the best computational method, the choice of a system of nuclear constants, and the specification of the errors in the results and of the possible range of variation of the design data. At the same time, the requirements on the accuracy of calculations are now becoming more stringent: In many cases it is necessary to determine the characteristics of radiation fields as closely as within 5%–15%, especially near the active zone of a reactor. The fourth stage in the development of shielding physics, which we have now entered, is thus a stage in which it is necessary to improve the quality of calculations. Part of this task is to find reliable estimates of the errors in the calculated results.

Another reason for determining and specifying the errors of calculations or measurements is to make it possible to compare the calculated and experimental data reported by different authors. In this regard it is interesting to note that a comparison of methods for measuring the integral characteristics of the field of neutrons and γ rays has shown[1] that the results of measurements carried out by different methods disagree by a factor of up to about 2, and this disagreement goes beyond the errors of the measurement methods estimated by the authors. A similar comparison of the errors of four benchmark experiments by different groups[2] has shown that in every case the discrepancies between the calculated and experimental results exceed the errors in the calculated data estimated by the authors. The discrepancy between the calculated and experimental data has averaged about 30%–70% for integral characteristics of the radiation field and a factor of 3 to 5 for differential characteristics. These examples demonstrate that in many cases the errors are not being estimated adequately.

It is obvious that further research on radiation shielding is important, has practical applications, and is needed now. It is pertinent to recall that today the thickness of the shielding for, say, a nuclear reactor is 3–3.5 m in terms of ordinary concrete, and the cost of the shielding in modern nuclear-engineering installations can range up to 20%–30% of the total cost. We can expect comparable figures for fusion power. Estimates put the cost of the shielding for a fusion-power installation at about 15% of the cost of the reactor.

A more or less generally accepted methodology has now been developed for solving radiation-theory problems for any nuclear-engineering installation. This methodology must include estimates of the computational errors.

The methodology for solving radiation-transport problems can be illustrated in its most general form as in Fig. 1. We can distinguish four basic directions of these studies:

(1) The development of methods, algorithms, and programs for radiation-shielding calculations.
(2) The acquisition, representation, and evaluation of data on the cross sections for the interaction of radiations with matter and the development of nuclear-data libraries, including libraries of multigroup constants, along with error files.
(3) Reference and parametric experiments.
(4) Study and determination of the errors and their sources in shielding calculations. A study of the sensitivity of the calculated results to the input data for the problem makes it possible to evaluate the overall computational approach and to develop a program for further work to refine the overall approach.

The approach used to study errors and their sources in the overall methodology of the solution of the radiation-transport problem makes it possible to incorporate the feedback shown by the dashed lines in Fig. 1.

The results of research in the first three of these directions have been reported in tens of Soviet and foreign monographs, handbooks, and hundreds of articles in the periodic literature.

Foremost among the computational methods are the discrete-ordinates method and the Monte Carlo method in its many modifications. Backed up by

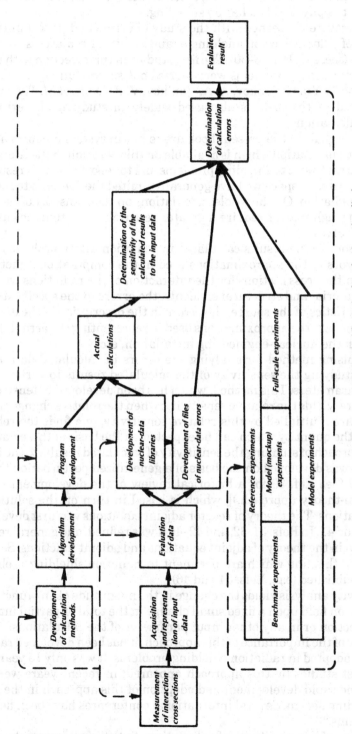

Figure 1. Overall methodology for solving problems in the physics of radiation shielding.

an appropriate organization of the programs, adequate computer power, and satisfactory libraries of nuclear data, these methods can solve most practical problems in the physics of radiation shielding.

The literature concerned with the study of errors and uncertainties in the results of calculations of shielding against ionizing radiations is rather more sparse (see, e.g., Refs. 3–30). The first studies in this direction with reference to reactor-design problems were carried out successfully by L. N. Usachev and also by Yu. G. Bobkov, A. A. Van'kov, V. A. Dulin et al. The experience acquired in this field is also used widely in studying the errors of shielding calculations.

The basic goals of the present book are to put in systematic form and to analyze that information which is available on this question in the literature and which is pertinent to shielding problems and to report research results of the authors. To save space we have generally limited the contents to sources of neutron radiation. On the whole, calculations on the transport of neutron radiation through matter require a greater research effort than do calculations for γ sources.

The errors in the results calculated in radiation-safety problems stem from the errors in the approximations used in the computational methods, the errors in the cross sections for the interaction of the radiations with the shielding material and structural elements, the errors in the specification of the characteristics of the source, the errors in the description of the detector response function, the approximations used in representing the actual shielding layout in the nuclear-engineering installation, etc.

A promising method for studying the errors in calculated data is the method of studying the sensitivity of the calculated results to variations in the input parameters. This method, which has been developed extensively in the literature, yields the relative sensitivity; when the relative changes in the input data are multiplied by this relative sensitivity, one finds the relative changes in the quantity which can be expected on the basis of these data.

The approach of studying the sensitivity of calculated results to the input parameters was developed on the basis of a generalized perturbation theory [Refs. 31–39 (See Ref. 39 for a historical review of the development of the perturbation-theory approach.)], which is based in turn on the solution of adjoint equations. The theory of reactor adjoint equations was first developed in Refs. 40 and 41. In Refs. 31, 32, and 42–45 it was refined, and general results were derived in the theory of adjoint equations and adjoint functions. Several studies along this line which are pertinent to radiation-shielding problems have been published (e.g., Refs. 34 and 46).

Sensitivity analysis yields the change in the magnitude of any functional of the radiation field upon a fixed small change in the total or partial interaction cross section or in any other input parameter of the calculations.

Because of the importance of this approach, it has been developed rapidly and rapidly adopted in radiation-shielding problems. It was only 15 years ago that the first studies by this approach appeared; in recent years we have witnessed the rapid development and adoption of this approach in the practice of shielding design. Several international conferences have been held on these problems.[3-6]

Questions of the sensitivity of calculations to the input parameters and estimates of computational errors have been discussed at recent international[7,8] and Soviet[9-11] conferences on shielding against ionizing radiations. While only a few of the reports at the First (1974; Ref. 9) and Second Soviet All-Union (1978; Ref. 10) Scientific Conferences on Ionizing-Radiation Shielding for Nuclear-Engineering Installations dealt with estimates of calculation errors and problems of the sensitivity of the calculated results to errors in the input data,[47-54] at the Third All-Union Conference (1981; Ref. 11) a special section was devoted to these questions. This approach should become a standard part of the procedure for planning and designing shielding.

This approach can solve many problems involving shielding against ionizing radiations. Among these problems are that of estimating the errors in shielding calculations which stem from various uncertainties and approximations in the input parameters (the cross sections for the interaction of the radiations with matter; the specification of the characteristics of the source, the shielding, and the detector response function; etc.); that of identifying the input parameter which is most important in terms of its effect on the results (e.g., that of determining the importance of some partial cross section or another, or of groups of these cross sections, on the results of calculations); that of generating recommendations on the accuracy required of measurements, calculations, and the processing of data on the cross sections for the interaction of radiations with matter; that of correcting the interaction cross sections to achieve the best fit with the set of data from benchmark experiments; that of studying the propagation of radiations through a medium; that of planning new reference model experiments which are sensitive to the parameters of interest; that of generating quantitative estimates of the importance of small changes in the design for the overall design of the shielding of nuclear-engineering installations; and that of generating practical recommendations for improving the quality of shielding design.

To illustrate the possibilities of this approach, we cite a single calculation result. Figure 2 shows the relative sensitivity of the neutron dose at the upper cap of the Clinch River Breeder Reactor (CRBR) to the total cross section for the interaction of neutrons with sodium.[22,55] The curve in Fig. 2 shows that the relative sensitivity reaches a very large value (10^3 per unit lethargy) at the minimum in the sodium interaction cross section at 300 keV. As a result, a change of, say, 10% in the cross section near the minimum will cause a change of 60% in the total dose at the upper cap. This example shows, in particular, the importance of an accurate knowledge of the interaction cross section at its minimum.

Examination of the sensitivity of the result to the energy dependence of the cross section for the interaction of the radiations with the matter, as in the example we just saw, can reveal so-called energy channels in which most of the radiation propagates. This concept might accordingly be called "energy channel theory." [14]

A spatial channel theory has also been proposed[23,25,26] for determining the primary channels in the shielding volume through which most of the radiation flows from the source to the detector.

The procedure of studying the sensitivity of calculated results to changes in the input parameters is thus the starting point not only for determining

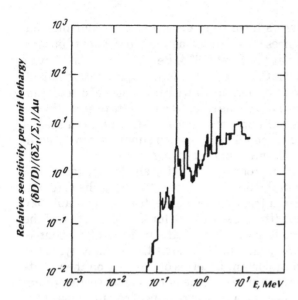

Figure 2. Relative sensitivity of the neutron dose at the upper cap of the Clinch River Breeder Reactor to the total neutron cross section of sodium.

the errors in shielding calculations but also for solving many of the independent problems of practical importance which were listed above. For these reasons, sensitivity analysis has a prominent place in this book.

It is fair to say that, in general, sensitivity analysis has furnished specialists a powerful tool for improving the quality of radiation-shielding calculations.

The very fact that specialists are turning to the problem of determining computational errors is evidence of the well-developed state of the computational approach for solving problems of radiation transport in matter. Only on this basis has it become possible to take up problems of determining computational errors.

On the other hand, it would be incorrect to say that the determination of shielding-calculation errors has been developed so thoroughly that all that remains to be done is a matter of computation techniques. The widespread dissemination of error computational methods in the solution of radiation-transport problems will still require much work, primarily the development and dissemination of convenient specialized program systems such as SWAN-LAKE (Ref. 15), ZAKAT (Refs. 30 and 56), FORSS (Refs. 5 and 57), and INDÉKS for calculations, combined with the existing, widely used programs for shielding calculations, with cross-section libraries generated from evaluated files of the errors in the radiation-matter interaction constants.

The first of the four chapters following this Introduction presents to the reader the topics covered in this book. Section 1.1 presents the basic concepts and terminology for the questions covered in this book, making the book self-contained. Also discussed here is the mathematical-statistics approach to a determination of the errors in calculated results (Sec. 1.3). The possibilities of sensitivity analysis are analyzed (Sec. 1.4). The authors have chosen to devote a separate section (1.2) to the physical interpretation of the relative-sensitivity function, which has played a significant role in the widespread adoption of this method in practice.

Chapter 2 discusses various methods for determining the relative sensitivity and the errors of calculations. Section 2.1 introduces the reader to the preparation of the constants required in radiation-transport calculations and the errors in the cross sections for the interaction of radiations with matter. Section 2.2 introduces the reader to programs for calculating the calculation errors and the relative sensitivity of the results to the input parameters. These sections deal with only the original program complexes and the system of constants which have been used for calculations by the authors of this book. It has been assumed that the reader can learn about other programs and systems from the corresponding publications. The systematic presentation of models for benchmark calculations (Sec. 2.3) gives the reader a better picture of the volume of research which has been carried out and makes it unnecessary to describe these models in detail in the subsequent chapters.

The third chapter is the main one. It discusses the methodological errors, calculating the errors and the relative sensitivity of the results to the input parameters for homogeneous, heterogeneous, and actual shielding layouts, primarily for one-dimensional geometries. The abundance of specific material on these questions rules out examination and analysis of all the studies in the literature in this short book. The authors have restricted discussion to the results of their own research and the most typical data of other authors. The emphasis is on the relationship between the calculation errors and the radiation-matter interaction cross section.

The errors in the input data (these are the errors in the interaction cross sections, the errors in the shielding layout, etc.) are usually comparatively small. The linear approximation is thus usually satisfactory for evaluating the errors in the calculated results within the errors in the input data. It is important to know the error of this approximation. The final section in Chap. 3 deals with estimates of nonlinear effects in sensitivity analysis.

Experiment is the criterion of truth. The final chapter of this book accordingly deals with the role played by basic experimental studies in problems of sensitivity analysis and the determination of computational errors.

The International System of Units (SI) is used in this book. For the convenience of the reader, units which do not conform to the System and which were previously in widespread use are given in square brackets along with the SI units and the units permitted by Soviet All-Union State Standard 8.417-81 in several complicated cases.

V. V. Bolyatko wrote Chap. 4; M. Yu. Vyrskii wrote Sec. 3.4; A. I. Ilyushkin wrote Sec. 3.5; V. P. Mashkovich wrote the Introduction and Secs. 1.1 and 1.4; V. K. Sakharov wrote Sec. 2.3; A. P. Suvorov wrote Sec. 3.1; G. N. Manturov and M. N. Nikolaev wrote Secs. 1.3 and 2.1; A. I. Ilyushkin and V. P. Mashkovich wrote Sec. 1.2; A. I. Ilyushkin and V. K. Sakharov wrote Sec. 3.2; G. N. Manturov, M. N. Nikolaev, and A. I. Ilyushkin wrote Sec. 2.2; and M. Yu. Vyrskii, G. N. Manturov, M. N. Nikolaev, A. I. Ilyushkin, and A. O. Suvorov wrote Sec. 3.3.

By no means do the authors believe that they are capable of producing a definitive book in this new field of shielding physics. However, they do hope that they are furnishing the reader a book which will promote the widespread adoption of the methods for determining calculation errors and the sensitivity of the calculated results to the input parameters in the practical solution of problems of radiation transport in media.

We are deeply indebted to our comrades in our work, especially Yu. I. Balashov, A. M. Zhezlov, V. I. Ivanov, I. N. Kachanov, V. A. Klimanov, I. I. Linga, A. A. Stroganov, V. A. Utkin, and A. V. Shikin for useful discussions and assistance in this work.

V. P. M.

Chapter 1
Basic concepts, definitions, and approaches

1. Characteristics of a radiation field, calculation errors, and sensitivity of calculated results to the input parameters

In this section we examine the basic concepts and terminology pertinent to the characteristics of a radiation field, to the errors of measurements and calculations, and to the sensitivity of calculated results to changes in the input parameters. An unambiguous interpretation of these concepts is required for a clear understanding of the material in this book.

Characteristics of a radiation field (Refs. 58–66)

A complete representation of a radiation field would require the specification of the number of particles, their energies, and the directions from which they arrive at any point in the medium at each time.

The most comprehensive information on a steady-state field at some point r_0 is given by the energy-angular flux density of the particles.

The energy-angular flux density of the particles at detection point r_0, $\varphi(r_0,E,\Omega)$, is the number dN of particles with energies between E and $E + dE$ which are propagating in the directions that are determined by the element of solid angle $d\Omega$ containing the given direction Ω and which, during the time interval dt, intersect the area element dS, which is centered at the fixed point r_0 of the field under consideration (the normal to this area element coincides with the selected radiation propagation direction Ω), divided by the area dS, the time inverval dt, the energy interval dE, and the element of solid angle, $d\Omega$:

$$\varphi(r_0,E,\Omega) = \frac{d^4N}{dS\, dt\, dE\, d\Omega}.$$

In the International System of Units (SI), the unit for the quantity $\varphi(r_0,E,\Omega)$ is $1/(m^2\ s\ J\ sr) = m^{-2}\ s^{-1}\ J^{-1}\ sr^{-1}$. A convenient unit in practice is $1/(cm^2\ s\ MeV\ sr) = cm^{-2}\ s^{-1}\ MeV^{-1}\ sr^{-1}$, in which the megaelectron volt, a multiple of a non-SI unit, is used; this non-SI unit is allowed by Soviet All-Union State Standard 8.417-81 (Ref. 59).

1

Less-detailed characteristics of a field are sufficient in many practical problems. In such cases, the following integral quantities of the energy-angular characteristics of the radiation field are used: the angular flux density of the particles,

$$\varphi(\mathbf{r}_0,\mathbf{\Omega}) = \frac{d^3N}{dS\,dt\,d\Omega} = \int_0^\infty \varphi(\mathbf{r}_0,E,\mathbf{\Omega})dE\;; \qquad (1.1)$$

the energy flux density of the particles,

$$\varphi(\mathbf{r}_0,E) = \frac{d^3N}{dS\,dt\,dE} = \int_{4\pi} \varphi(\mathbf{r}_0,E,\mathbf{\Omega})d\Omega\;; \qquad (1.2)$$

and the particle flux density

$$\varphi(\mathbf{r}_0) = \frac{d^2N}{dS\,dt} = \int_{4\pi} \varphi(\mathbf{r}_0,\mathbf{\Omega})d\Omega = \int_0^\infty \varphi(\mathbf{r}_0,E)dE\;. \qquad (1.3)$$

The units for the quantities defined in (1.1)–(1.3) can be written easily, since they are ultimately found through an integration of the quantity $\varphi(\mathbf{r}_0,E,\mathbf{\Omega})$, for which the units were given above.

The solution of many problems requires information on characteristics of the field which are determined by particles whose energy E lies above a certain boundary E_b. It then becomes possible, for example, to determine the flux density $\varphi(\mathbf{r}_0,E>E_b)$ of particles with energies above E_b at a fixed detection point $\mathbf{r}_0$ from the relation

$$\varphi(\mathbf{r}_0,E>E_b) = \int_{E_b}^\infty \varphi(\mathbf{r}_0,E)dE\;. \qquad (1.4)$$

Those characteristics of the field whose arguments are the energy E and the direction in which the particle is moving ($\mathbf{\Omega}$) are called "differential characteristics." In this sense, the characteristics $\varphi(\mathbf{r}_0,E,\mathbf{\Omega})$, $\varphi(\mathbf{r}_0,\mathbf{\Omega})$, and $\varphi(\mathbf{r}_0,E)$ are differential characteristics. Characteristics which do not depend on E or $\mathbf{\Omega}$ are "integral characteristics." For brevity, the words "differential" and "integral" are usually omitted since the argument of a quantity clearly shows which characteristic is being discussed.

The differential and integral particle flux densities which we have introduced here may vary in space, in time, or in both. In the most general case of a variation in both space and time, we arrive at the concept of a space-time energy-angular particle flux density $\varphi(\mathbf{r},t,E,\mathbf{\Omega})$.

For the flux characteristics $\varphi(\mathbf{r}_0,E)$ and $\varphi(\mathbf{r}_0)$ of the field, all the particles are typically given an equal weight, regardless of the direction in which they are moving.

Current characteristics of the radiation field are convenient in many problems.

The most comprehensive information on a steady-state field at a fixed detection point $\mathbf{r}_0$ is given by the energy-angular particle current density $\mathbf{J}(\mathbf{r}_0,E,\mathbf{\Omega})$, which is equal in modulus to $\varphi(\mathbf{r}_0,E,\mathbf{\Omega})$ but differs from the latter in being a vector, whose direction is the particle propagation direction $\mathbf{\Omega}$.

We thus have

$$\mathbf{J}(\mathbf{r}_0,E,\Omega) = \Omega\varphi(\mathbf{r}_0,E,\Omega) . \qquad (1.5)$$

By analogy with Eqs. (1.1)–(1.3) we can write the following integral quantities of the energy-angular characteristic of the particle current density at the point $\mathbf{r}_0$:

$$\mathbf{J}(\mathbf{r}_0,\Omega) = \int_0^\infty \mathbf{J}(\mathbf{r}_0,E,\Omega)dE = \int_0^\infty \Omega\varphi(\mathbf{r}_0,E,\Omega)dE = \Omega\varphi(\mathbf{r}_0,\Omega) ; \qquad (1.6)$$

$$\mathbf{J}(\mathbf{r}_0,E) = \int_{4\pi} \mathbf{J}(\mathbf{r}_0,E,\Omega)d\Omega = \int_{4\pi} \Omega\varphi(\mathbf{r}_0,E,\Omega)d\Omega ; \qquad (1.7)$$

$$\mathbf{J}(\mathbf{r}_0) = \int_{4\pi} \mathbf{J}(\mathbf{r}_0,\Omega)d\Omega = \int_{4\pi} \Omega\varphi(\mathbf{r}_0,\Omega)d\Omega . \qquad (1.8)$$

The integration of the angular current characteristics over solid angle in these formulas reduces to a summation of the vectors by the rules of vector algebra. The directions of the resultant vectors $\mathbf{J}(\mathbf{r}_0,E)$ and $\mathbf{J}(\mathbf{r}_0)$ are generally not known at the outset. They can be determined simply from the functions $\varphi(\mathbf{r}_0,E,\Omega)$ and $\varphi(\mathbf{r}_0,\Omega)$. A resultant vector determines the directions of the current across a unit area perpendicular to the direction of the current.

In practice, we are usually not interested in the current across an area whose position is strictly fixed by the resultant current vector but the current across an area which is spatially oriented by a condition of the problem. We define the spatial orientation of this area by means of the unit vector $\mathbf{k}$, which is perpendicular to the surface of the area. The problem formulated here is equivalent to determining the projection of the energy-angular current density of particles at the point $\mathbf{r}_0$ in the direction specified by the vector $\mathbf{k}$, i.e., $J_k(\mathbf{r}_0,E,\Omega)$. This projection can be found from the expression

$$J_k(\mathbf{r}_0,E,\Omega) = \mathbf{J}(\mathbf{r}_0,E,\Omega)\cdot\mathbf{k} = (\Omega\cdot\mathbf{k})\varphi(\mathbf{r}_0,E,\Omega) = \varphi(\mathbf{r}_0,E,\Omega)\cos(\hat{\Omega}\cdot\hat{\mathbf{k}}) . \quad (1.9)$$

In a calculation of $J_k(\mathbf{r}_0,E,\Omega)$, each particle is therefore taken into account along with a corresponding weight, which depends on the angle between the vectors Ω and $\mathbf{k}$. Specifically, it is equal to $\cos(\hat{\Omega}\cdot\hat{\mathbf{k}})$.

By analogy with Eqs. (1.1)–(1.3) and (1.6)–(1.8), we could write the following integrated values of the projection of the energy–angular current density of particles in the direction specified by the vector $\mathbf{k}$:

$$J_k(\mathbf{r}_0,\Omega) = \mathbf{J}(\mathbf{r}_0,\Omega)\cdot\mathbf{k} = \varphi(\mathbf{r}_0,\Omega)(\Omega\cdot\mathbf{k}) = \varphi(\mathbf{r}_0,\Omega)\cos(\hat{\Omega}\cdot\hat{\mathbf{k}}) ; \qquad (1.10)$$

$$J_k(\mathbf{r}_0,E) = \mathbf{J}(\mathbf{r}_0,E)\cdot\mathbf{k} = \int_{4\pi} \varphi(\mathbf{r}_0,E,\Omega)(\Omega\cdot\mathbf{k})d\Omega = \int_{4\pi} \varphi(\mathbf{r}_0,E,\Omega)\cos(\hat{\Omega}\cdot\hat{\mathbf{k}})d\Omega ;$$

$$(1.11)$$

$$J_k(\mathbf{r}_0) = \mathbf{J}(\mathbf{r}_0)\cdot\mathbf{k} = \int_{4\pi} \varphi(\mathbf{r}_0,\Omega)(\Omega\cdot\mathbf{k})d\Omega = \int_{4\pi} \varphi(\mathbf{r}_0,\Omega)\cos(\hat{\Omega}\cdot\hat{\mathbf{k}})d\Omega . \quad (1.12)$$

For the current characteristics of the particle field we can also use a dependence on time and/or space.

The units of the current characteristics of the field are the same as the corresponding units of the flux characteristics, but it must be kept in mind that physically the current and flux quantities are quite different. (Ref. 63).

A time dependence, space dependence, or space and time dependence of the energy-angular energy flux density of the particles, of the energy current density of the particles, or of the energy current density of the particles onto the direction of the vector **k**, which runs perpendicular to some arbitrary area S_k, or their functionals is also widely used as a characteristic of radiation fields. These concepts are introduced by analogy with the definitions above; there is the difference here that these concepts, in all cases, characterize not the number of particles but the energy which they transport.

Dose characteristics of the radiation field are frequently used to evaluate the effect of radiation in practical problems.

Let us examine the absorbed and equivalent doses, which are quantities used in the present book.

The absorbed radiation dose (or, simply, radiation dose) D is the ratio of the average energy (dW) which is transferred by ionizing radiation to the matter in a volume element to the mass (dm) of the matter in this volume element:

$$D = dW/dm .\qquad(1.13)$$

The unit of absorbed dose in the SI is the joule per kilogram (J/kg). A joule per kilogram corresponds to the absorption of an energy of 1 J of any kind of ionizing radiation in 1 kg of the irradiated material. This unit is called the gray (Gy).

A unit of absorbed dose which does not conform to the International System is the rad. One rad corresponds to the absorption of 100 erg of energy of any kind of ionizing radiation in 1 g of irradiated material.

Thus 1 rad = 100 erg/g = 1×10^{-2} Gy; 1 Gy = 1 J/kg = 100 rad.

The concept of a dose equivalent D_{equ} is introduced to estimate the biological effect of radiation of arbitrary composition.

For radiation of a single quality, the equivalent dose D_{equ} is the product of the absorbed radiation dose in the biological tissue, D, and the quality factor K of this radiation in the given element of biological tissue:

$$D_{equ} = DKf .\qquad(1.14)$$

The quality factor K determines the dependence of the unfavorable biological consequences of the irradiation of a human in small doses on the total linear energy transfer of the radiation. This is a regulated value of the relative biological effectiveness of radiation, established for monitoring the extent of radiation danger during chronic irradiation. Values of K are given in Refs. 58 and 61–63, among other places.

The SI unit of equivalent dose is the sievert (Sv). One sievert is equal to one gray divided by the quality factor K. The sievert is a unit of equivalent dose of any type of ionizing radiation in biological tissue which has the same biological effect as the absorbed dose in 1 Gy of standard x or γ radiation.

A unit of equivalent dose which does not conform to the SI is the rem. One rem is equal to one rad divided by the quality factor K.

We thus have 1 rem = 1×10^{-2} Sv; 1 Sv = 100 rem.

Table I. Values of the maximum ceiling doses equivalent δ_{equ} and the absorbed doses δ per unit fluence used in the calculations.

Energy group	Range of energy group, MeV	$\delta, \mu Gy\ m^2/n$ $\left[\dfrac{rad\ cm^2}{n}\right]$	$\delta_{equ}, \mu Sv\ m^2/n$ $\left[\dfrac{rem\ cm^2}{n}\right]$
1	10.5–6.5	6.7×10^{-9}	3.8×10^{-8}
2	6.5–4	5.6×10^{-9}	3.9×10^{-8}
3	4–2.5	4.6×10^{-9}	3.7×10^{-8}
4	2.5–1.4	4.2×10^{-9}	3.7×10^{-8}
5	1.4–0.8	3.7×10^{-9}	3.7×10^{-8}
6	0.8–0.4	2.8×10^{-9}	2.7×10^{-8}
7	0.4–0.2	1.8×10^{-9}	1.8×10^{-8}
8	0.2–0.1	1.3×10^{-9}	1.0×10^{-8}
9	0.1–0.0465	9.0×10^{-10}	7.0×10^{-9}
10	4.65×10^{-2}–2.15×10^{-2}	6.7×10^{-10}	3.4×10^{-9}
11	2.15×10^{-2}–1.0×10^{-2}	5.7×10^{-10}	2.8×10^{-9}
12	1.0×10^{-2}–4.65×10^{-3}	5.6×10^{-10}	1.4×10^{-9}
13	4.65×10^{-3}–2.15×10^{-3}	5.9×10^{-10}	1.5×10^{-9}
14	2.15×10^{-3}–1.0×10^{-3}	6.0×10^{-10}	1.5×10^{-9}
15	1.0×10^{-3}–4.65×10^{-4}	6.0×10^{-10}	1.5×10^{-9}
16	4.65×10^{-4}–2.15×10^{-4}	6.2×10^{-10}	1.6×10^{-9}
17	2.15×10^{-4}–1.0×10^{-4}	6.6×10^{-10}	1.6×10^{-9}
18	1.0×10^{-4}–4.65×10^{-5}	6.7×10^{-10}	1.7×10^{-9}
19	4.65×10^{-5}–2.15×10^{-5}	6.4×10^{-10}	1.6×10^{-9}
20	2.15×10^{-5}–1.0×10^{-5}	6.1×10^{-10}	1.5×10^{-9}
21	1.0×10^{-5}–4.65×10^{-6}	5.7×10^{-10}	1.4×10^{-9}
22	4.65×10^{-6}–2.15×10^{-6}	5.1×10^{-10}	1.3×10^{-9}
23	2.15×10^{-6}–1.0×10^{-6}	4.7×10^{-10}	1.2×10^{-9}
24	1.0×10^{-6}–4.65×10^{-7}	3.8×10^{-10}	0.95×10^{-9}
25	4.65×10^{-7}–2.15×10^{-7}	3.4×10^{-10}	0.85×10^{-9}
26	2.15×10^{-7}–2.15×10^{-9}	3.2×10^{-10}	1.0×10^{-9}

In tissue, the radiation dose is distributed nonuniformly among the biological objects. The maximum in the dose distribution may be observed in the interior of the tissue rather than at its surface.

A customary measure of the extent of the effect of irradiation is the ceiling value of the irradiation dose in the human body. The use of ceiling values of tissue doses rules out the possibility that the tolerable dose will be exceeded at any point in the human body. The word "ceiling" is usually omitted for brevity. We will conform to this usage here, understanding "tissue dose" as meaning the ceiling value.

The ceiling equivalent dose δ_{equ} and the ceiling absorbed dose δ per unit fluence used in calculations by the authors of this book are listed in Table I. Strictly speaking, the units $\mu Gy\ m^2/n$ (n = neutron), rad cm^2/n, $\mu Sv\ m^2/n$, and rem cm^2/n used in Table I should be replaced by $\mu Gy\ m^2$, rad cm^2, $\mu Sv\ m^2$, and rem cm^2, respectively, without the specification of "neutron." Such specifications should be attached to the name of the quantity; in these cases, e.g., we should say an "absorbed neutron dose per unit fluence of $6.7 \times 10^{-9}\ \mu Gy\ m^2$." However, here and below, we will deviate from this rule in several cases to clarify the data which we are presenting.

The absorbed dose rate (P) or the equivalent dose rate (P_{equ}) is the ratio of the increment in the absorbed dose (dD) or in the equivalent dose (dD_{equ}) over the time interval dt to this time interval:

$$P = dD/dt; \quad P_{equ} = dD_{equ}/dt. \tag{1.15}$$

The units for these quantities are the units of absorbed or equivalent dose (or their multiples or submultiples) divided by the appropriate unit of time.

Errors in measurements or calculations (Refs. 67–73)

The scientific literature unfortunately lacks an unambiguous terminology in the branch of metrology concerned with the accuracy of measurements. For example, such terms as "error" (oshibka), "accuracy" (tochnost'), and "uncertainty" (neopredelennost') are frequently used to characterize errors (pogreshnost). To illustrate the ambiguity, we can cite a few examples from various papers on radiation transport:

(1) "The dependence of the integral current albedo on r and d is described with an accuracy (tochnost') $\pm 10\%...$."

(2) "The accuracy (tochnost') of the observed values of the ^{7}Be yields is $\pm 13\%$."

(3) "The errors (oshibki) in the measurement of the bremsstrahlung yields lie between the limits 6–9%."

(4) "The error (oshibka) in the calculation of the dose after the generation of 4000 case histories is 5%."

(5) "The error (pogreshnost') in the determination of the relative concentration which is associated with the statistical errors (oshibki) of the calculation is no more than 3% on the average."

According to Soviet All-Union State Standard 16263-70 (Ref. 67), the terms "error" (oshibka) and "accuracy" (tochnost') should be replaced by the term "error" (pogreshnost') in all these examples.

We turn now to the definitions of the basic terms pertinent to the characterization of the quality of measurements or calculations. For the most part, we are following All-Union State Standard 16263-70 (Ref. 67) in the determination of the error of measurements.

Error in the result of a measurement or calculation (measurement or calculation error)

A characteristic of the result of a measurement or calculation which is the deviation of the value found for the quantity from its actual value (in a determination of the error of measurements) or from its exact value (in a determination of the error of calculations). A few comments are in order regarding this definition.

The "true" or "exact" value means that value of a physical quantity which we are striving to find in accordance with the problem posed in the experiment or calculation, respectively, and which would reflect, qualitatively and quantitatively, the corresponding property of the object.

Imperfections of the procedure and conditions of measurements or calculations in the problem posed have the consequence in practice that the result of the measurement or calculation deviates from the true or exact value, respectively, which remains unknown. In practice, therefore, we can find only an estimate of the true or exact value. For experimental studies, the estimate of the true value is called the "actual value" of the physical quantity. The actual value of a physical quantity (or simply "the actual value") is a value of the physical quantity which is so close to the true value that it can be used instead of the latter for the given problem. The introduction of the concept of a true or exact value is nevertheless necessary, since without it one would find it difficult to determine the error of the result of a measurement or calculation. It follows from this discussion that in practice we can find only an approximate estimate of the error of the result of a measurement or calculation; in other words, the errors of a measurement or calculations are also approximate quantities.

A distinction is made between "absolute" and "relative" errors in a measurement calculation.

Absolute error of the result of a measurement or calculation (absolute error of a measurement or calculation)
The error of a measurement or calculation expressed in units of the quantity being measured.

Relative error of the result of a measurement or calculation (relative error of a measurement or calculation)
The ratio of the absolute error of the measurement or calculation to the true value of the measured quantity or the exact value of the calculated quantity. This quantity can be expressed as a percentage.

Here are the major components of the error in a result.

Random component of the error of the result of a measurement or calculation (random error)
A component of the error in the result of a measurement or calculation of a specific physical quantity which varies in a random way when the measurements or calculations of the given quantity are repeated. A random error is of a statistical nature, obeys the laws governing random quantities, and is estimated by the method of mathematical statistics. Such errors are accordingly also called "statistical errors." Examples of random errors are the error in the reading of the display of a measurement instrument, the error due to the random law of radioactive decay in a determination of the activity of a sample, and the statistical errors in the calculation of radiation transport by the Monte Carlo method.

Systematic component of the error of the result of a measurement or calculation (systematic error)
A component of the error in the result of a measurement or calculation of a specific physical quantity which remains constant or varies in a regular way when the measurements or calculations of the same quantity are repeated. Examples of systematic errors are a shift of the scale of an instrument, an incorrect calibration, an error due to a constant decrease in the working current in the circuit of an electrical measurement instrument, and the methodological error in a calculation on neutron transport which results from the

use of a finite number of terms in the expansion of a function of the scattering cross section in a series in Legendre polynomials. Attempts are usually made to eliminate systematic errors from the result of experiments or calculations. One is then left with only the uncorrected residue of systematic error. When these operations are carried out, it is customary to use the concepts of a correction or correction factor.

Correction

A value of a physical quantity, of the same nature as that being measured or calculated, which should be algebraically summed with the result of the measurement or calculation in order to eliminate a systematic component of the error.

Correction factor

A number by which the result of the measurement or calculation should be multiplied in order to eliminate a systematic component of the error.

Coarse error in the result of a measurement or calculation (coarse error)

A component of the error in the result of a measurement or calculation of a specific physical quantity which is substantially greater than the error expected under the given conditions. This error component is sometimes called a "miss" (promakhi) or "errors" (oshibki). Examples of coarse errors are errors which result from the entry of incorrect input data, from a malfunction of a computer in the course of a calculation, or from the appearance of unexpected strong influences on a measurement.

Small errors of all types, both systematic and random, correspond to a high accuracy. It is for this reason that one must not identify the terms "error" and "accuracy" with each other.

Accuracy of a measurement or calculation

A characteristic of the quality of a measurement or calculation which reflects the proximity of the results to the true value of the measured quantity or the exact value of the calculated quantity.

Quantitatively, the accuracy can be expressed as the reciprocal of the modulus of the relative error. For example, if the error is $10^{-2}\% = 10^{-4}$, the accuracy is 10^4.

All-Union State Standard 8.011-72 (Ref. 68) establishes the following indices of the accuracy of measurements: the interval within which the measurement error lies, with a given probability; the interval within which the systematic component of the error of measurements lies, with a given probability; numerical characteristics of the systematic component of the error of a measurement; numerical characteristics of the random component of the error of a measurement; the distribution function (probability density) of the systematic component of the error of a measurement; the distribution function (probability density) of the random component of the error of a measurement.

In practice it is generally not necessary to specify simultaneously all of these accuracy indices in each measurement. There are various ways to express the accuracy of a measurement, the most common being to specify the interval within which the total error of the measurement lies, within an established probability.

Sensitivity of the results of calculations to the input parameters (Refs. 16, 35, 74, 75)

In 1961, Moorhead[75] introduced the concept of coefficients of the sensitivity to changes in neutron nuclear data in the breeding problem of a fast-neutron reactor. He used the resulting sensitivity coefficients to determine the permissible errors in nuclear data, which result in a specified accuracy in the calculation of the breeding ratio.

Soviet scientists have developed an effective method for calculating sensitivity coefficients for various reactor parameters by means of a generalized perturbation theory.[31-39]

We denote by R a linear functional of the radiation field, e.g., the particle flux density, the fluence of particle energy, the dose rate, the heat evolution, or the radiation damage.

We assume that the quantity R depends on N different input parameters $X_1, X_2, ..., X_N$; i.e., $R = R(X_1, X_2, ... X_N)$. In radiation shielding problems, these parameters might be the total or partial cross sections for the interaction of radiations with matter, the parameters of the functions specifying the source or the response of a detector, geometric characteristics, etc.

Let us assume that one of the possible parameters X_n changes by a small amount ΔX_n. This change will result in a change ΔR in the result.

The limit of the ratio of $\Delta R/R$ to $\Delta X_n/X_n$ as $\Delta X_n/X_n$ approaches zero is then called the "relative sensitivity of the functional of the radiation field R to the parameter X_n," $p_{Rn}(X_1, X_2, ..., X_N)$:

$$p_{Rn}(X_1, X_2, ..., X_N) = \lim_{\Delta X_n/X_n \to 0} \frac{\Delta R/R}{\Delta X_n/X_n}, \tag{1.16}$$

where $n = 1, 2, ..., N$.

In the literature, the quantity p_{Rn} is also frequently called the "sensitivity function," the "sensitivity profile," or the "sensitivity coefficient."

The value of the function $p_{Rn}(X_1, X_2, ..., X_N)$ may thus be interpreted as the percentage change in R upon a 1% change in X_n.

Let us assume that the parameters $X_1, X_2, ..., X_N$ have acquired independent small increments $\Delta X_1, \Delta X_2, ..., \Delta X_N$. The relative change in the radiation-field functional R can then be determined from the following formula, with an accuracy to small quantities of second order in $\Delta X_n/X_n$:

$$\frac{\Delta R}{R} \simeq \sum_{n=1}^{N} p_{Rn} \frac{\Delta X_n}{X_n}. \tag{1.17}$$

If $\delta X_1, \delta X_2, ..., \delta X_N$ are small errors within which $X_1, X_2, ..., X_N$ are known, the relative error R can be determined with the help of expression (1.17):

$$\frac{\overline{\delta R}}{R} = \sqrt{\left(\frac{\delta R}{R}\right)^2} \simeq \sqrt{\sum_{n,n'=1}^{N} p_{Rn} p_{Rn'} \frac{\delta X_n}{X_n} \frac{\delta X_{n'}}{X_{n'}} \rho_{nn'}^*}$$

$$= \sqrt{\sum_{n,n'=1}^{N} p_{Rn} p_{Rn'} w_{nn'}}, \tag{1.18}$$

where $\rho_{nn'}^*$ are the correlation coefficients of the quantities $\delta \overline{X}_n, \delta \overline{X}_{n'}$; $w_{nn'}$ is an element of the covariation matrix $\widehat{W}$; and $\widehat{W} = \|w_{nn'}\|_{NN}$.

The relations given above were derived under the assumption that $\Delta R/R$ is a linear function of $\Delta X_n/X_n$.

The relative-sensitivity function can be used to deal with nonlinear effects.

Working from Eq. (1.17), we can easily write an expression for the differential of $\ln R$ as a function of the variables $X_1, X_2, ..., X_N$:

$$d \ln R = \frac{dR}{R} = \sum_{n=1}^{N} p_{Rn}(X_1, X_2, ..., X_N) \frac{dX_n}{X_n}. \qquad (1.19)$$

Let us assume that the value of R is known for some values of the parameters. Let us assume, for example, that for the values $X_1^*, X_2^*, ..., X_N^*$ we have $R^* = R(X_1^*, X_2^*, ..., X_N^*)$. Integrating expression (1.19) along an arbitrary path connecting the points $(X_1^*, X_2^*, ..., X_N^*)$ and $(X_1, X_2, ..., X_N)$, we then find[74]

$$R(X_1, X_2, ..., X_N) = R^* \exp\left(\int_{(X_1^*, X_2^*, ..., X_N^*)}^{(X_1, X_2, ..., X_N)} \sum_{n=1}^{N} p_{Rn}(X_1', X_2', ..., X_N') \frac{dX_n'}{X_n'} \right). \qquad (1.20)$$

The exact expression for $\Delta R/R$ becomes

$$\frac{\Delta R}{R} = \exp\left(\int_{(X_1^*, X_2^*, ..., X_N^*)}^{(X_1 + \Delta X_1, X_2 + \Delta X_2, ..., X_N + \Delta X_N)} \sum_{n=1}^{N} p_{Rn}(X_1', X_2', ..., X_N') \frac{dX_n'}{X_n'} \right) - 1. \qquad (1.21)$$

Expression (1.21) can be used to determine the relative error:

$$\frac{\delta R}{R} = \sqrt{\overline{\left(\frac{\Delta R}{R}\right)^2}}$$

$$= \left[1 - 2 \exp\left(\overline{\int_{(\overline{X}_1, \overline{X}_2, ..., \overline{X}_N)}^{(\overline{X}_1 + \Delta X_1, \overline{X}_2 + \Delta X_2, ..., \overline{X}_N + \Delta X_N)} \sum_{n=1}^{N} p_{Rn}(X_1', X_2', ..., X_N') \frac{dX_n'}{X_n'}} \right) \right.$$

$$\left. + \exp\left(\overline{2 \int_{(\overline{X}_1, \overline{X}_2, ..., \overline{X}_N)}^{(\overline{X}_1 + \Delta X_1, \overline{X}_2 + \Delta X_2, ..., \overline{X}_N + \Delta X_N)} \sum_{n=1}^{N} p_{Rn}(X_1', X_2', ..., X_N') \frac{dX_n'}{X_n'}} \right) \right]^{1/2}, \qquad (1.22)$$

where $\overline{X}_n$ is the mathematical expectation of X_n.

Expressions (1.17) and (1.18) can be derived from Eqs. (1.21) and (1.22). Assuming $p_{R_n} = \text{const}$ in some region $[X_1 - \Delta X_1, X_1 + \Delta X_1] \times \cdots \times [X_N - \Delta X_N, X_N + \Delta X_N]$, expanding $\Delta R/R$ and $\delta R/R$ in Taylor series around $X_1, X_2, ..., X_N$, and retaining small quantities of up to second order in $\Delta X_n/X_n$, we find expressions (1.17) and (1.18).

Incidentally, expression (1.20) can be used for an approximate description of the analytic dependence of R on $X_1, X_2, ..., X_N$ in certain ranges of these parameters from the calculated values of R^* and p_{Rn} in the construction of approximate calculation methods.

That aspect of the sensitivity analysis which involves calculations of the increment ΔR and the relative error $\overline{\delta R}/R$ has been labeled "predictive" analysis.

2. Physical interpretation of the relative sensitivity

Let us examine the physical interpretation of relative sensitivity,[16,35] which has played an important role and which has been largely responsible for the widespread adoption of sensitivity analysis in practical calculations on radiation shielding.

For simplicity we write the transport equation and the result of a calculation in the form

$$\left.\begin{array}{l} \hat{L}\varphi = S; \\ \varphi|_{r\in\Gamma} = 0 \quad \text{for} \quad \Omega n < 0; \\ R = \langle q, \varphi \rangle. \end{array}\right\} \tag{1.23}$$

The angle brackets in (1.23) mean an integration over the entire phase space ξ (over the spatial variable r, the angular variable Ω, and the energy variable E), φ (ξ) is the spatial energy-angular particle flux density, $S(\xi)$ is the source specification function, $\hat{L}$ is the operator of the kinetic equation, n is the outward normal to the surface Γ which is the boundary of the shielding, and $q(\xi)$ is the detector response function.

We assume that the input parameters $X_1, X_2, ..., X_N$ have acquired small, independent increments $\Delta X_1, \Delta X_2, ..., \Delta X_N$. The increment ΔR in the functional R can then be written as follows according to linear perturbation theory[16,35]:

$$\Delta R = \langle \Delta q, \varphi \rangle + \langle \Delta S, \varphi^* \rangle - \langle \varphi, \Delta L^* \varphi^* \rangle, \tag{1.24}$$

where φ^* is the solution of the adjoint problem

$$\left.\begin{array}{l} \hat{L}^*\varphi^* = q; \\ \varphi^*|_{r\in\Gamma} = 0 \quad \text{for} \quad \Omega n > 0. \end{array}\right\} \tag{1.25}$$

Here L^* is the adjoint operator for the Lagrangian:

$$(\hat{L}\varphi, \varphi^*) = (\varphi, \hat{L}^*\varphi^*). \tag{1.26}$$

We assume that the parameters $X_1, X_2, ..., X_N$ appear linearly in the quantities q, S, and L^*, i.e.,

$$\left.\begin{array}{l} q = \sum_n q(X_n, \xi) = \sum_n X_n q_n(\xi); \\ S = \sum_n S(X_n, \xi) = \sum_n X_n S_n(\xi); \\ \hat{L}^* = \sum_n \hat{L}^*_{X_n} = \sum_n X_n \hat{L}^*_n. \end{array}\right\} \tag{1.27}$$

We then have

$$\left.\begin{array}{l} \Delta q = \sum_n \Delta X_n q_n(\xi); \\ \Delta S = \sum_n \Delta X_n S_n(\xi); \\ \Delta \hat{L}^* = \sum_n \Delta X_n L^*_n. \end{array}\right\} \tag{1.28}$$

Substituting Eq. (1.28) into Eq. (1.24), multiplying the resulting expression by $(1/R)/(\Delta X_n/X_n)$, and taking the limit $\delta = \Delta X_n/X_n \to 0$, we find the following expression for the relative sensitivity (R) to the input parameter X_n:

$$p_{Rn}(X_1,X_2,...,X_N) = \frac{\langle q(X_n),\varphi \rangle}{R} + \frac{\langle S(X_n),\varphi^* \rangle}{R} - \frac{\langle \varphi, \widehat{L}^*_{X_n} \varphi^* \rangle}{R} . \qquad (1.29)$$

To explain the physical interpretation of $p_{Rn}(X_1,...,X_N)$, we consider separately each of the terms in expression (1.29). The first term is the part of R which is due to the detection of the particles by a detector with an efficiency $q(X_n,\xi)$. The two other terms contain the importance (or danger) function φ^*. The value of φ^* at the point ξ is known to be equal to the contribution to R made by a particle which is injected into phase space at this point. The product of $\varphi^*(\xi)$ and the number of particles which are "injected" (or produced) at ξ is equal to the contribution of these particles to the functional R. The second term in expression (1.29) is thus equal to the part of R which is due to the particles of the source $S(X_n,\xi)$, i.e., those particles for whose production the parameter X_n "is responsible." The quantity $\varphi \widehat{L}^*_{X_n}$ in the third term is interpreted as the source of particles which are produced as the result of the interaction processes described by the operator $\widehat{L}^*_{X_n}$. The third term in Eq. (1.29) is thus equal to the part of R which is determined by the interaction processes or by other characteristics described by the operator $\widehat{L}^*_{X_n}$, i.e., characterized by the parameter X_n.

The quantity $p_{Rn}(X_1,X_2,...,X_N)$ is thus the part of R which is determined by the processes characterized by the parameter X_n. The physical interpretation of $p_{Rn}(X_1,X_2,...,X_n)$ explains the name of this quantity, "the relative sensitivity of R to the input parameter X_n."

To explain this interpretation of the relative sensitivity of the results of the calculations to the input parameters, we consider several examples.

The energy dependence of the relative sensitivity of R to the detector response function is a qualitative characteristic of the part of R which is determined by the detection parameter for particles with energy E:

$$p_{Rq}(E) = \frac{1}{R} \int_V \int_\Omega q(\mathbf{r},E,\Omega)\varphi(\mathbf{r},E,\Omega) dV \, d\Omega . \qquad (1.30)$$

The relative sensitivity of R to the parameters of the source specification function in shielding volume V_m is a quantitative characteristic of the part of R which is due to the particles of the source which are produced in the volume V_m:

$$p_{RS}(V_m) = \frac{1}{R} \int_{V_m} \int_\Omega \int_E S(\mathbf{r},E,\Omega)\varphi^*(\mathbf{r},E,\Omega) dV \, d\Omega \, dE . \qquad (1.31)$$

The energy dependence of the relative sensitivity of R to the elastic-scattering cross section in shielding volume V_m is a quantitative characteristic of that part of R which is governed by the parameter of the process of elastic scattering of particles with an energy E in the volume V_m:

$$p_{R,\text{el}}(V_m,E) = \frac{1}{R} \int_{V_m} \int_{\Omega} \varphi(\mathbf{r},E,\Omega)\bigg(-\Sigma_{\text{el}}(\mathbf{r},E)\varphi^*(\mathbf{r},E,\Omega)$$

$$+ \int_{E'} \int_{\Omega'} \Sigma_{\text{el}}(\mathbf{r},E\to E',\Omega\to\Omega')\varphi^*(\mathbf{r}E',\Omega')dE'\,d\Omega')dE'\,d\Omega'\bigg)dV\,d\Omega ,$$

$$(1.32)$$

where $\Sigma_{\text{el}}(\mathbf{r},E\to E',\Omega\to\Omega')$ is the doubly differential cross section for elastic scattering, and

$$\Sigma_{\text{el}}(r,E) = \int_{E'} \int_{\Omega'} \Sigma_{\text{el}}(\mathbf{r},E\to E',\Omega\to\Omega')dE'\,d\Omega'.$$

In practice, studies of the energy dependence of the relative sensitivity of R to the radiation-matter interaction cross sections have been most common. As was mentioned in the Introduction, such studies make it possible to distinguish energy channels in which the particles interact effectively during transport from the source to the detector.

A useful and important supplement to energy channel theory is spatial channel theory,[23,25,26] for distinguishing spatial channels along which particles which dominate R propagate from the source to the detector.

Solving problems of spatial channel theory involves determining the "response current"

$$\mathbf{R}(\mathbf{r}) = \int_E \int_\Omega \Omega\varphi(\mathbf{r},E,\Omega)\varphi^*(\mathbf{r},E,\Omega)dE\,d\Omega , \qquad (1.33)$$

which characterizes the contribution to the reading (response) of the detector for the given detected effect from the current of particles which pass through an area at the spatial point $\mathbf{r}$ between the source and the detector.

The quantity in (1.33) was introduced in Refs. 23, 25, and 26, where it was called the "contribution current."

If the source and detector of the radiation are spatially separated, and if the shielding is bordered by a vacuum, the response current $\mathbf{R}(\mathbf{r})$ is related to the calculated radiation-field functional R by

$$R = \left| \int_\Gamma \mathbf{R}(\mathbf{r})\mathbf{n}\,d\Gamma \right| . \qquad (1.34)$$

Here $\mathbf{n}$ is the normal to the surface Γ, and the integral is evaluated over the surface Γ, which divides the shielding into two parts, one containing only the source, and the other containing only the detector.

By sequentially making cuts with a set of such surfaces, $\Gamma_1,\Gamma_2,...,\Gamma_M$, and choosing for each cut the points $\mathbf{r}_1,\mathbf{r}_2,...,\mathbf{r}_M$ for which the modulus of the response current takes on its maximum values, one can determine the spatial channels for inleakage of the radiation to the detector.

A generalization of the channel theories[76] is an expression for determining the spatial-energy particle-transport channels which basically determine R:

$$\mathbf{R}(\mathbf{r},E) = \int_\Omega \Omega\varphi(\mathbf{r},E,\Omega)\varphi^*(\mathbf{r},E,\Omega)d\Omega . \qquad (1.35)$$

By analogy with the concepts of a particle current and a particle flux, one introduces the response flux

$$C(\mathbf{r}) = \int_E \int_\Omega \varphi(\mathbf{r},E,\Omega)\varphi^*(\mathbf{r},E,\Omega)dE\,d\Omega\,, \tag{1.36}$$

which has no obvious relationship with the contribution to the calculated result R.

It was shown in Ref. 25 that in many cases the relation $C(\mathbf{r}) = |R(\mathbf{r})|$ holds.

3. Mathematical-statistics approach to the determination of the error of a calculated result

As mentioned earlier, the result of a calculation of the radiation-field functionals R at the point $\mathbf{r}$ can be determined as linear functionals of the energy-angular radiation flux density $\varphi(\mathbf{r},E,\Omega)$:

$$R = \int q(\mathbf{r},E,\Omega)\varphi(\mathbf{r},E,\Omega)dV\,dE\,d\Omega\,, \tag{1.37}$$

where the detector response function q characterizes the sensitivity of the particular radiation-field functional R in which we are interested to the radiation flux density. In turn, the spatial energy-angular radiation flux density $\varphi(\mathbf{r},E,\Omega)$ is determined by radiation transport equation (1.23).

The error in the calculated prediction of radiation-field functionals R is determined by the following factors: First, there is the error in the detector response function $q(\mathbf{r},E,\Omega)$. For most functionals, the spatial and energy-angular variables of the detector response function can be separated:

$$q(\mathbf{r},E,\Omega) = K(\mathbf{r})F(E,\Omega) \quad\text{or even}\quad q(\mathbf{r},E,\Omega) = K(\mathbf{r})T(E)O(\Omega)\,.$$

Most frequently, the energy dependence of the response function $T(E)$ is known with the largest error. Second, the error in a calculated prediction is determined by the errors in the calculation of the radiation flux density $\varphi(\mathbf{r},E,\Omega)$:

$$\varphi(\mathbf{r},E,\Omega) = \hat{L}^{-1}S(\mathbf{r},E,\Omega)\,. \tag{1.38}$$

We can distinguish five components of the error in the determination of $\varphi(\mathbf{r},E,\Omega)$. The first component, which we will call the "calculation-constants component,*" results from an uncertainty in the specification of the characteristics of the interaction of the radiation with the shielding material. The second stems from the idealization of the actual, usually three-dimensional, geometry of the shielding which is unavoidable in calculations; the idealization frequently involves replacing the actual geometry by a two-dimensional or even one-dimensional geometric model. The third component is generated by the approximate description of the spatial-angular dependence of the radiation flux which is used in numerical methods for solving the transport equation. The fourth component stems from the approximate nature of the methods used to solve the transport equation (a methodological error). The fifth is determined by the errors in the specification of the characteristics of the external source S.

---

*A better translation might be "cross-section component".

These sources of errors in a calculated prediction can be put in three groups. The first includes the uncertainties which stem from the formulation of the problem and which are therefore impossible in principle to eliminate; examples are the uncertainties in the specification of the external source. In certain problems, the range of permissible variations in the characteristics of the source, e.g., its power, can be extremely broad. In other cases, e.g., in a calculation of the radiation shielding for the core of a nuclear reactor of specified power, the source can in principle be specified essentially exactly, but in real shielding calculations the source of the primary radiation must frequently be specified approximately. The second group contains the methodological errors, which can be lowered to an acceptable level if sufficiently accurate computation methods and a suitable computational procedure are used in the calculations. The third group contains the errors due to the constants (nuclear data).

The methodological and constant-related errors differ not only in nature but also in their properties. The methodological errors which stem from several approximations are of a systematic nature, while for the constant-related errors there is typically a randomness, since these errors are generated by a set of different factors. Foremost among these factors are the inexact specification of the nuclear data (interaction cross sections, energy-angular distributions of the scattered radiation, etc.). Another source of constant-related error is the technology used to fabricate the shield or, more precisely, the tolerances on the average shielding density and the composition of the shielding materials, because of which the actual nuclear concentrations are different from the design concentrations.

Since the nuclear-physics and technological components of the constant-related error are statistical in nature, an estimate of their contribution to the error in the calculation of radiation-field functionals in shielding is naturally carried out by the methods of mathematical statistics.

Let us outline some typical problems which can be solved by these methods. We denote by $\underline{Z} = \{z_n\}$ $(n = 1,2,...,N_1)$ the column vector of nuclear concentrations (specified by the designer) in various zones of the shield, and we denote by $\underline{Y} = \{y_n\}$ $(n = N_1 + 1, N_1 + 2,...,N_1 + N_2)$ the column vector of the cross sections for the interaction of radiation with the shielding materials. These are the cross sections which are used in the calculations. For shielding calculations in the multigroup approximation, $\underline{Y}$ is understood as a set of group constants.* These constants are of course different from the real values $\underline{Y}'' = \{y_n''\}$, which we do not know. These differences stem from differences between the design nuclear concentrations and the actual concentrations $\underline{Z}'' = \{z_n''\}$ in the shielding and from the approximation of the actual particle flux density in the shielding in the derivation of the group constants $\underline{Y}$. The deviations of z_n from z_n'' and of y_n from y_n'' are random in nature and usually small (a few percent), and the distributions of these deviations can be approximated as normal distributions obeying

$$\underline{P}(\underline{Z}|\underline{Z}'') = (2\pi)^{-N_1/2}(\det \widehat{\underline{V}})^{-1/2} \exp\left[-\tfrac{1}{2}(\underline{Z}'' - \underline{Z})^T \widehat{\underline{V}}^{-1}(\underline{Z}'' - \underline{Z})\right] ; \quad (1.39)$$

---

*A better translation might be "group cross section".

$$\underline{P}(\underline{Y}|\underline{Y}'') = 2\pi^{-N_2/2}(\det \widehat{W})^{-1/2} \exp\left[-\tfrac{1}{2}(\underline{Y}''-\underline{Y})^T\widehat{W}^{-1}(\underline{Y}''-\underline{Y})\right],\ (1.40)$$

where $\underline{P}(\underline{Z}|\underline{Z}'')$ is the probability density that the nuclear concentrations in the actual shield will have the values $\underline{Z}''$ under the condition that the design specifies values $\underline{Z}$ and $\underline{P}(\underline{Y}|\underline{Y}'')$ is the probability density that the set of group constants corresponding to the real shield are $\underline{Y}''$, while the set $\underline{Y}$ is used in the calculation. The superscript T in Eqs. (1.39) and (1.40) means transposition; det $\widehat{V}$ and det $\widehat{W}$ are the determinants of the matrices $\widehat{V}$ and $\widehat{W}$, respectively.

The values of the concentrations $\underline{Z}$ and of the constants $\underline{Y}$ adopted during the design are the mathematical expectations of the random quantities $\underline{Z}''$ and $\underline{Y}''$, respectively:

$$\underline{Z} = \overline{\underline{Z}}'' = \int \underline{Z}'' \underline{P}(\underline{Z}|\underline{Z}'')d\underline{Z}'' ;\qquad (1.41)$$

$$\underline{Y} = \overline{\underline{Y}}'' \equiv \int \underline{Y}'' \underline{P}(\underline{P}|\underline{Y}'')d\underline{Y}'' .\qquad (1.42)$$

In (1.39) and (1.40), $\widehat{V}$ is the covariance matrix in the errors of the nuclear concentrations, and $\widehat{W}$ is the covariance matrix of the errors in the constants:

$$\widehat{V} = \|v_{nn'}\|;\qquad \widehat{W} = \|w_{nn'}\|;$$

$$v_{nn'} = (\overline{z_n - z_n''})(z_{n'} - z_{n'}'') \equiv \int (z_n - z_n'')(z_{n'} - z_{n'}'')\underline{P}(\underline{Z}|\underline{Z}'')d\underline{Z}'' ;\quad (1.43)$$

$$w_{nn'} = (\overline{y_n - y_n''})(y_{n'} - y_{n'}'') \equiv \int (y_n - y_n'')(y_{n'} - y_{n'}'')\underline{P}(\underline{Y}|\underline{Y}'')d\underline{Y}'' .\quad (1.44)$$

The diagonal elements of the matrices $\widehat{V}$ and $\widehat{W}$ are the variances of the corresponding concentrations or constants; the off-diagonal elements are the covariances of the errors, which are zero if these errors are statistically independent. The covariance matrices of the errors are necessary characteristics of the data which are used, and we will assume that they are given.

We now denote by $\widetilde{R} = \{\widetilde{R}_k\}$ $(k = 1,2,...,K)$ the set of functionals to be calculated. We denote by $\widehat{G}$ and $\widehat{H}$ the rectangular matrices of the derivatives of these functionals with respect to the nuclear concentrations and group constants:

$$\widetilde{G} \equiv \left\|\frac{\partial \widetilde{R}_k}{\partial z_n}\right\|;\qquad \widehat{H} \equiv \left\|\frac{\partial \widetilde{R}_k}{\partial y_n}\right\| .\qquad (1.45)$$

In practice, the elements of these matrices can be calculated by perturbation theory, e.g., by the ZAKAT program (Sec. 2.2).

Let us estimate the accuracy of the calculated predictions. If the deviations of z_n and y_n from z_n'' and y_n'' are not too large, the relationship between the errors in $\mathbf{Z}$ and $\mathbf{Y}$ and the errors which they cause in $\widetilde{R}$ can be assumed to be linear:

$$\widetilde{R} - \widetilde{R}'' = \widehat{G}(\underline{Z} - \underline{Z}'') + \widehat{H}(\underline{Y} - \underline{Y}'') .\qquad (1.46)$$

where $\widetilde{R} = \{\widetilde{R}_k''\}$ is the vector of real values of the calculated functionals. In this case the probability density that the actual values of the functionals correspond to the set $\tilde{R}''$, while their calculated values are $\tilde{R}$, is described by a normal distribution:

$$P(\tilde{R}|\tilde{R}'') = (2\pi)^{-K/2}(\det \hat{U})^{-1/2} \exp[-\tfrac{1}{2}(\tilde{R}'' - \tilde{R})^T \hat{U}^{-1}(\tilde{R}'' - \tilde{R})], \quad (1.47)$$

where $\hat{U}$ is the covariance matrix of the errors in the functionals of interest:

$$\hat{U} = \hat{G}\hat{V}\hat{G}^T + \hat{H}\hat{W}\hat{H}^T. \quad (1.48)$$

This expression can be used directly to estimate the calculation-constants components of the errors in the calculation of the characteristics of radiation shielding.

We now consider ways to improve the accuracy of calculated predictions by incorporating the data from integral (macroscopic) benchmark experiments. If the calculation-constants components of the errors in Eq. (1.48) are too large, we can attempt to reduce them if we take the data from benchmark experiments into account. We assume that we have the results of measurements of a set of integral characteristics $R^0 = \{R_n^0\}$ $(n = 1,2,3,...,J)$ of the radiation fields. We assume that the conditions in these experiments were such that, with the computation facilities available, it is possible to calculate the measured characteristics so accurately that the methodological errors in the results of the calculations are much smaller than the errors in the experimental determination of these characteristics and can be ignored. We assume that the covariance matrix $\hat{M}$ of the experimental errors is known (it is determined by the experimental conditions). The results of the calculations of the measured quantities are functionally related to the constants Y used in the calculations. We denote these results by $R = R(Y)$. Since the constants Y differ from the exact values Y'', the calculated results will also differ from the actual values R'', which would be found if we used the exact constants Y'' in the calculations:

$$R'' = R''(Y''). \quad (1.49)$$

The latter equality emphasizes the fact that both the methodological errors and the errors within which the dimensions and compositions of the experimental installations are known are unimportant for benchmark experiments (this is one of the bases for choosing these experiments as benchmark experiments).

The values of the experimentally measured characteristics R^0 of course also differ from the real values R''; the corresponding probability density is

$$P(R|R'') = (2\pi)^{-J/2}(\det \hat{M})^{-1/2} \exp[-\tfrac{1}{2}(R'' - R)^T\hat{M}^{-1}(R'' - R)]. \quad (1.50)$$

We thus have two groups of experiments: microscopic experiments, whose results are used to construct the constants Y which we use in the calculations, and whose deviations from the true values are described by distribution (1.40); and macroscopic or integral experiments, which also carry information on the true values of the constants Y''. Incorporating the latter type of information should lead us to different—refined—estimates of the constants. We denote these estimates by Y' and their covariance matrix by $\hat{W}'$. To find refined estimates, we turn to the principle of maximum likelihood, which can be outlined as follows.

We construct the probability density that, for the real values of the constants Y'' in the macroscopic and microscopic experiments, we will find precisely those values (R^0 and $\underline{Y}$, respectively) which were actually found. Since the errors of these groups of experiments are completely independent, this probability density is

$$\underline{P}(\underline{R}^0,\underline{Y}|\underline{Y}'') = \underline{P}[\underline{R}^0|\underline{R}(\underline{Y}'')]\underline{P}(\underline{Y}|\underline{Y}''),\tag{1.51}$$

where $\underline{P}[\underline{R}^0|\underline{R}(\underline{Y}'')]$ and $\underline{P}(\underline{Y}|\underline{Y}'')$ are determined in accordance with Eq. (1.50) and Eqs. (1.39) and (1.40). The quantity $\underline{P}(\underline{R}^0,\underline{Y}|\underline{Y}'')$, thought of as a function of Y'' with given R^0 and $\underline{Y}$, is called the "likelihood function." The principle of maximum likelihood recommends that the estimate $\underline{Y}'$ be chosen in such a manner that the likelihood function reaches a maximum under the condition $Y'' = Y'$. Such an estimate is called the "maximally likely estimate." Maximally likely estimates are competent, effective, and asymptotically normal; if the likelihood function is sufficiently regular, they have the minimum possible variances.[77] In the case in which we are interested here, the latter condition holds: Since the errors in the constants are small, we can set

$$\underline{R}(\underline{Y}') = \underline{R}(\underline{Y}) + \widehat{F}(\underline{Y}' - \underline{Y}),\tag{1.52}$$

where $\widehat{F}$ is the rectangular matrix of derivatives of the characteristics being calculated with respect to the constants:

$$\widehat{F} \equiv \left\| \frac{\partial R_j^0}{\partial y_n} \right\|.\tag{1.53}$$

In the linear approximation, Eq. (1.52), the following quadratic form appears in the argument of the exponential function of likelihood function (1.51):

$$-\tfrac{1}{2}\left\{[\underline{R} - \underline{R}^0 + \widehat{F}(\underline{Y}' - \underline{Y})]^T\widehat{M}^{-1}[\underline{R} - \underline{R}^0 + \widehat{F}(\underline{Y}' - \underline{Y})]\right.$$
$$\left. + (\underline{Y}' - \underline{Y})^T\widehat{W}^{-1}(\underline{Y}' - \underline{Y})\right\}.\tag{1.54}$$

The problem of seeking the maximum of the likelihood function reduces in this case to a search for the minimum of this quadratic form. Differentiating (1.54) with respect to each of the constants $\underline{Y}'$, and equating the derivatives to zero, we find a system of linear equations in the refinements of the constants:

$$(\underline{W}^{-1} + \widehat{F}^T\widehat{M}^{-1}\widehat{F})(\underline{Y}' - \underline{Y}) = \widehat{F}^T\widehat{M}^{-1}(\underline{R} - \underline{R}^0).\tag{1.55}$$

Its solution is

$$\underline{Y}' - \underline{Y} = (\widehat{W}^{-1} + \widehat{F}^T\widehat{M}^{-1}\widehat{F})^{-1}\widehat{F}^T\widehat{M}^{-1}(\underline{R} - \underline{R}^0).\tag{1.56}$$

Expression (1.56) requires inverting matrices of rank N_2. It is possible, however, to derive the alternative expression

$$\underline{Y}' - \underline{Y} = \widehat{W}\widehat{F}^T(\widehat{M} + \widehat{F}\widehat{W}\widehat{F}^T)^{-1}(\underline{R} - \underline{R}^0).\tag{1.57}$$

When this expression is used, it is sufficient to invert a matrix of rank J to find the solution. Since the number of quantities which are measured in benchmark experiments is substantially smaller than the number of constants, it is preferable to use Eq. (1.57).

The covariance matrix of the errors of the refined constants, $\widehat{W}'$, is

$$\widehat{W}' = (\widehat{W}^{-1} + \widehat{F}^T\widehat{M}^{-1}\widehat{F})^{-1} = \widehat{W} - \widehat{W}\widehat{F}^T(\widehat{M} + \widehat{F}\widehat{M}\widehat{F}^T)^{-1}\widehat{F}\widehat{W}.\tag{1.58}$$

The use of the refined constants Y' to calculate the characteristics of the shield being designed naturally improves the accuracy of the calculated predictions of the characteristics $\tilde{R}$ in which we are interested. Replacing $\hat{W}$ and $\hat{W'}$ in (1.48), we find

$$\hat{U'} = \hat{G}\hat{V}\hat{G}^T + \hat{H}\hat{W'}\hat{H}^T = \hat{U} - \hat{H}\hat{W}\hat{F}^T(\hat{M} + \hat{F}\hat{W}\hat{F}^T)^{-1}\hat{F}\hat{W}\hat{H}^T . \quad (1.59)$$

How informative are benchmark experiments? We assume that of all the quantities $\tilde{R}$ which are calculated for the design process we are particularly interested in one of them, $\tilde{R}_k$. We see from (1.59) that by taking into account the set of benchmark experiments we reduce the variance of the calculated estimate of this quantity by

$$I_a(\tilde{R}_k) = u_{kk} - u'_{kk} = \sum_{n}^{N_2} \sum_{n'}^{N_2} p_{R_k n} \, \Delta w_{nn'} p_{R_k n'} , \quad (1.60)$$

where $\Delta w_{nn'}$ are the elements of the matrix $\Delta \hat{W} = \hat{W}\hat{F}^T(\hat{M} + \hat{F}\hat{W}\hat{F}^T)^{-1}\hat{F}\hat{W}$. It was suggested in Ref. 39 that this quantity be called the "decrease in the informativeness of the set of macroscopic experiments with respect to the characteristic $\tilde{R}_k$."

This definition differs from that used in information theory, where the quantity which is used as a measure of information is the reciprocal of the variance, so that the informativeness would be defined as $(u'_{kk})^{-1} - (u_{kk})^{-1}$. Definition (1.60), however, is convenient from both the computational standpoint and the standpoint of practical interpretation. The concept of a relative informativeness was introduced in Ref. 78:

$$I_r(\tilde{R}_k) = I_a(\tilde{R}_k)/u_{kk} . \quad (1.61)$$

This relative informativeness is convenient for comparing the informativeness of a given set of benchmark experiments with respect to various characteristics. By evaluating the informativeness of the results expected for several alternative versions of an experimental program, one can choose the best version.

It can be seen from the expression for $\Delta \hat{W}$ that no matter how accurate the benchmark experiment (i.e., no matter how small the elements of the matrix $\hat{M}$), its informativeness cannot exceed a limit which is determined, on the one hand, by the accuracy with which the constants are known (i.e., by the matrix $\hat{W}$) and, on the other, by the extent to which the sensitivities to the constants of the quantities measured in the benchmark experiments (the matrix $\hat{F}$) are similar to the sensitivities in the design characteristic of interest (the row matrix $\hat{H}_k$).

These questions are discussed more thoroughly in Refs. 39 and 77–80.

4. Capabilities of sensitivity analysis

Let us consider several problems which can be solved by means of relative-sensitivity functions. We will illustrate the solutions of these problems with some simple examples.

(1) Estimating the error of calculations due to uncertainties (small er-

rors) in the input parameters. To solve this problem we need information on the relative sensitivity, the errors in the input data, and the correlation coefficients. Among the error components, one of particular practical importance is the component which stems from the errors in the cross sections for the interaction of the radiations with matter (a calculation-constants component).

A study of the sensitivity to the cross sections makes it possible, in particular, to determine the effect of some partial cross section or other, or group of such cross sections, on the results of the calculations.

At present, linear perturbation theory is the only effective tool which we have for estimating the calculation-constants component of the error in shielding calculations. However, in each case it is necessary to justify the use of linear perturbation theory in these estimates.

Let us illustrate the possibilities of sensitivity analysis for the particular case of determining the error in the calculation of the neutron and γ field at a distance of 2 km from a thermonuclear radiation source in an infinite medium (air).[81] A special role is played in the formation of the relative sensitivity in this problem by the partial cross section for the (n,α) reaction with nitrogen. The importance of the $^{14}N(n,\alpha)$ reaction is determined in particular by the minimum in the cross section for this reaction near an energy of 5 MeV.

Estimates show that the error in the total tissue dose in air is 14%, with a confidence interval of 0.67. Previous estimates which ignored correlation led to an error of 30%.

The errors in the cross sections for the reaction $^{14}N(n,\alpha)$ are responsible for 45% of the total error in the tissue dose. This example illustrates the importance of considering correlations in the determination of an error.

Errors in the specification of the characteristics of the source, the shielding, and the detector response also play an important role in a determination of the errors of calculations. The sensitivity of results to the source parameters makes it possible, for example, to determine the effect of some region or other of the source specification function on the result of the calculations. For the source described by an energy-group representation we can determine the importance and the role of the ith group of the source by calculating the contribution of this energy group to the result of the calculations of R.

(2) Estimates of the calculation error which stems from the approximations or limitations of the method are very important in solving transport-theory problems.

For example, it is important to resolve the question of the errors which stem from the representation of the scattering characteristic by a finite number of terms when it is expanded in a series in Legendre polynomials. This example is illustrated by Table II, which shows those relative errors in the calculation of the flux density of fast neutrons with energies $E \gtrsim 1.4$ MeV, $\varphi_F(E \gtrsim 1.4$ MeV), and the equivalent dose rate P_{equ} behind a spherical iron shield 2 m in diameter, with an isotropic point source of ^{252}Cf at the center of the sphere, which result from the switch from the P_5 approximation of the scattering characteristic to lower-order approximations. The data in this table show that in this example the flux density of fast neutrons with $E \gtrsim 1.4$ MeV can be calculated within an error of 3% in the P_2 approximation, while

Table II. Relative changes $\Delta\varphi_F(E\geqslant1.4$ MeV$)/\varphi_F(E\geqslant1.4$ MeV$)$ and $\Delta P_{equ}/P_{equ}$ due to the switch from the P_5 approximation for the scattering characteristic of lower-order approximations P_L (%).

Functional	P_0	P_1	P_2	P_3
$\Delta\varphi_F(E\geqslant1.4$ MeV$)/\varphi_F(E\geqslant1.4$ MeV$)$	-267	-50	-3	-0.2
$\Delta P_{equ}/P_{equ}$	-14	$+0.03$	—	—

P_{equ} can be calculated within an error of 0.03% in the P_1 approximation.

(3) A quantitative estimate of the importance to the design of the shielding for nuclear-engineering installations of small changes in the design. This estimate makes it possible to generate recommendations on the tolerable changes in the design.

Let us assume, for example, that an absorbed dose rate P is produced behind an iron barrier with a thickness $d = 50$ cm for a plane, unidirectional source of neutrons whose energy distribution is a fission spectrum at energies above 1.4 MeV and a $1/E$ spectrum at lower energies. We wish to determine how much we can change the thickness of the iron barrier without changing the absorbed dose rate by more than $\pm 30\%$.

We can solve this problem by making use of the value of $p_{P\Sigma_t}$, the relative sensitivity to the total neutron-iron cross section, Σ_t. The quantity $p_{P\Sigma_t}$ determined by perturbation theory from the dose rate turns out to be -1.83. As we noted above, the quantity $p_{P\Sigma_t}$ is numerically equal to the percentage change in the dose rate upon a 1% change in the total cross section; this is equivalent to a 1% change in the density of the medium, the mass, or the thickness. We thus find

$$\frac{\Delta P}{P} = p_{P\Sigma_t}\frac{\Delta\Sigma_t}{\Sigma_t} = p_{P\Sigma_t}\frac{\Delta d}{d}$$

and then

$$\Delta d = \frac{1}{p_{P\Sigma_t}}\frac{\Delta P}{P}d = \left(\frac{1}{-1.83}\right)(\pm 0.3)50 = \mp 8.2 \text{ cm.}$$

(4) Correcting the cross sections for the interaction of radiations with matter, primarily the group cross sections, is, from the mathematical standpoint, an optimization problem. However, because of the importance and practical significance of these input data in the solution of radiation-transport problems, the question of their optimization has emerged as an independent topic.[30]

"Correction of constants" can be connected with choosing the optimum way for partitioning the constants into groups along the energy scale. Sensitivity analysis shows that, for example, in a calculation of the radiation field at large distances from a source and for a source of high-energy neutrons, the switch from a 28-group representation to a 49-group representation makes it possible to significantly improve the accuracy of calculations of the characteristics of a radiation field.[82]

(5) Generation of recommendations on the necessary accuracy of measurements and of the processing of the raw data on the radiation-matter in-

Table III. Requirements on the accuracy (%) with which radiation-field functionals must be known in the shields of power and research reactors (Ref. 83).

Functional of field	Responses of various experts on the permissible errors for reactors							
	Power					Research		
Heat evolution from γ radiation in structural materials:								
(a) Without fissile materials	± 5	± 5	± 20	—	20	—	± 10	10
(b) With fissile materials	20	20	—	—	—	—	± 10	—
Heat evolution from γ radiation:								
(a) In the installation	± 10	± 10	20	± 10	—	20	± 10	—
(b) In the heat shield	—	—	20	—	20	—	—	—
(c) In the biological shield	—	—	—	—	30	—	—	10
Heat evolution in the control rods and shield	± 10	± 10	—	± 10	—	30	—	—
Concentration of displacements (in the course of radiation damage)								
(a) In the collector	± 15	± 50	± 10	—	—	30	± 20	—
(b) Near the reflector	± 15	± 50	± 10	—	—	—	± 20	—
Concentration of displacements (in the course of radiation damage)								
(a) In load-bearing structures	—	—	—	—	40	—	± 20	—
(b) Inside the shield system	± 25	± (50–100)	—	—	30	30	—	—
Activation								
(a) In steam generator	± 50	factor of 2–3	—	—	30	50	—	—
(b) In cooling pipes	—	—	—	± 25	30	—	—	—
(c) In turbine	—	—	—	—	50	—	—	—
Dose rate beyond the biological shield	factor of 2–3	factor of 2–3	factor of 2–3	± 50	40	factor of 2–3	± 50	10

teraction cross sections. The accuracy with which interaction cross sections must be known is determined by the accuracy with which we need to know the characteristics of the field in some specific problem or other. Different functionals of the characteristics of the radiation field naturally require different determination accuracies. Interesting in this regard is Ref. 83, which gives 17 answers from specialists from three countries on the accuracy with which the radiation-field functionals must be known in the shields for power and research reactors of various types and in irradiation installations. Some of this information is reproduced in Table III. Information of this sort can be used in sensitivity analysis to solve the problem which is the inverse of the problem of estimating the error of calculations and to determine the accuracy with which radiation-matter interaction cross sections must be known in order to calculate functionals of the selected type within the given error.

(6) Planning of reference experiments, primarily benchmark experi-

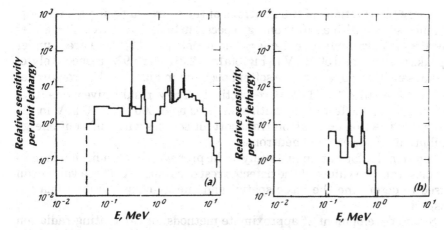

Figure 3. Relative sensitivity of the readings of a detector with a constant efficiency for detecting neutrons over the energy range 40–350 keV behind a layer of sodium 5 m thick to the total cross section for the interaction of neutrons with sodium. (a) Plane, isotropic fission source; (b) fission source at energies below 0.9 MeV.

ments, which are sensitive to those parameters of shielding calculations which have been singled out as most critical. This planning is intended for justifying and optimizing experimental studies for testing calculations and for refining and correcting raw data on radiation-matter interaction cross sections. The planning problem consists of choosing the simplest realization of the studies and the most informative model of the experiment.

In the planning of this problem, the relative sensitivity of the calculation to the parameter of interest must be made as large as possible in modulus in the experiments being planned. Associated with the planning of benchmark experiments is the problem of choosing data on the basis of computational studies and correcting the raw data from the most suitable experiments which have already been carried out.

(7) Study of the propagation of radiation in matter based on a physical interpretation of the relative sensitivity function $p_{Rn}(X_n)$. Sensitivity-analysis calculations may involve a qualitative and quantitative analysis of the results on the formation of radiation fields in a medium. The values of p_{Rn} (X_n) which are largest in modulus correspond to the processes which are the most important from the standpoint of the formation of the field: detection, generation, and the interaction of the radiation with the matter. An estimate of the effects of various interaction processes on the errors of calculations makes it possible to determine the least significant interaction processes.

We consider a small example pertaining to a fast-neutron nuclear reactor above whose core there is a layer of molten sodium 5 m thick. Figure 3 (Ref. 84) shows the energy dependence of the relative sensitivity of the readings of a detector with a constant efficiency for detection of the neutron flux density over the energy range 40–350 keV which is placed behind a 5-m-thick layer of sodium to the total neutron-sodium interaction cross section for this problem, for the case of a plane, isotropic source. These calculations were carried out by the SWANLAKE + ANISN programs with the ENDF/ BIV nuclear-data library for two energy distributions of the neutrons of the source: a fission source [Fig. 3(a)] and a source with a softer energy spectrum [a fission source at energies below 0.9 MeV, without higher-energy neutrons; Fig. 3(b)]. The areas under the peaks near 300 and 500 keV in Fig. 3(a) are 2.3% and 5.9%, respectively, of the total sensitivity $p_{\varphi \Sigma_t}$. For the softer spectrum in Fig. 3(b), the areas under these peaks are 61.4% and 8.9%, respectively. These results give a quantitative picture of the effect of the antiresonances near 300 and 500 keV in the sodium cross section on neutron transport for sources with different energy distributions of the emitted neutrons.

Additional information in a study of the propagation of radiation in matter comes from the values of the detector response current. These values can be used to determine the basic spatial and energy channels for radiation transport.

(8) The development of approximate methods for calculating radiation fields in media and studies of their range of applicability. This work is based on a study of the physics of particle transport through matter. Sensitivity analysis makes it possible to construct approximate functional relations and procedures and to evaluate the corresponding errors.

(9) The optimization of biological shielding. This is a necessary step in the design of a complex shield. The calculated relative sensitivities not only make it possible to carry out the laborious optimization process automatically but also can assist in the development of approximate methods in the initial stages of the optimization and in the evaluation of the errors in the calculations of radiation fields which stem from the use of these approximations. The optimization of a shield usually requires much computer time and is accordingly carried out by approximate methods.

To illustrate the solution of such problems we might cite Ref. 85, where use was made of the relative sensitivity of a functional to the thickness of the zones of heterogeneous medium to solve a neutron-converter problem. The optimum thicknesses were found for the layers of a four-layer converter, made of iron and graphite, with a total thickness of 30 cm, for intermediate-energy neutrons. The neutron energy distribution at the output from the converter was to approximate a particular model spectrum corresponding to fast-neutron reactors. The energy distribution was characterized by an average energy and by a buildup factor, defined as the ratio of the number of neutrons with energies below 1.4 MeV to the number with energies above 1.4 MeV.

As we mentioned above, sensitivity analysis, which makes it possible to solve these problems, includes in its calculation facilities some additional relationships which improve the quality of the calculations and which make it possible to take an iterative approach to the solution of a problem.

One possible flow chart for such an approach is shown in Fig. 4; the analysis below pertains to this particular flow chart. The problem-formulation stage is completed once the calculation model, the raw data on the radiation-matter interaction cross section, and the layout of the shield have been chosen for the given problem.

A radiation shield is usually a complex installation. The experience and physical intuition of the designer frequently make it possible to choose a satisfactory version of a shield which does not require an optimization. Opti-

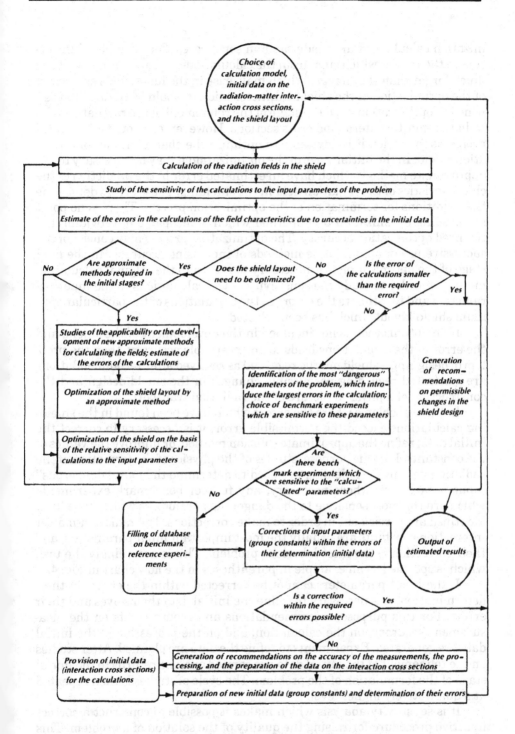

Figure 4. Flowchart for an iterative solution of a radiation-shielding problem.

mization calculations are inadvisable in some cases. For example, if the errors in the results which stem from the calculation-constants component are much larger than the changes which may arise in the functionals as a result of the optimization of the shield characteristics, it would be natural to postpone the optimization until the stage of the refinement of the raw data on the radiation-matter interaction cross sections. However, one could cite several problems for which it is advisable to optimize the characteristics of a radiation shield. In the initial stage of the optimization it may be necessary to use approximate methods which have an estimated error in the prediction of the characteristics of the radiation field in the shielding. These errors determine the conditions for termination of the optimization process: There is no point in pursuing optimization by a method which is incapable of providing the required optimization accuracy. The optimization process may consist of the successive use of approximate methods of increasing accuracy. In the final stage of the optimization it is necessary to return to the original calculation method in order to study the sensitivity of the calculations to the input parameters and to estimate the errors in the calculations of the particular optimum shield layout which has been selected.

If an optimization is not included in the development of the shield, and the error of the calculations leads to an acceptable accuracy, then the problem of designing a shield may be regarded as resolved after recommendations are generated regarding permissible changes in the shield design and the construction of a physical picture of the radiation transport.

In those problems where the errors which have been found in the shielding calculations exceed the permissible error, it is necessary to correct the initial data, refine the approximate solution method, or improve the values of the constants. For this purpose, studies of the physics of the formation of the radiation field in the shielding are used to determine the most "dangerous" input parameters; where necessary, any further benchmark experiments which are the most sensitive to the dangerous parameters which have been identified are carried out in order to make corrections. The radiation-matter interaction constants are usually the most important input parameters, and their correction is the most laborious problem. This is accordingly the task which is specified as an example in parentheses in the flow chart in Fig. 4.

If the input parameters cannot be corrected within the errors in their determination, it is necessary to refine the initial data themselves and their errors. For this purpose, recommendations and requirements on the measurement accuracy, on the calculation, and on the processing of the initial data are generated. Experiments to refine the data are planned. After studies on the refinement of the initial data are carried out, one returns to the initial stage of the formulation of the problem. The cycle of studies is then repeated at a qualitatively higher level.

It is sensitivity analysis which makes it possible to construct a correct iterative procedure for raising the quality of the solution of a problem. This procedure organically combines calculations and experiments, the generation of constants, and sensitivity analysis.

This example thus demonstrates how the possibilities of sensitivity analysis, which we looked at earlier in this section, can be used in a multifaceted way to solve shielding-calculation problems.

Chapter 2
Calculation methods and models

1. Preparation of cross sections for calculations; errors in interaction cross sections

For calculations on ionizing-radiation fields in media on a BÉSM-6 computer or a computer of the ES series, the ARAMAKO-2F system for generating cross sections can be recommended for preparing macroscopic and blocked microscopic cross sections ("ARAMAKO" is an acronym formed from the Russian words for "automated calculation of macroscopic constants").[86] This system differs from the earlier program complexes in its module organization, which significantly extends the capabilities of the system.

The ARAMAKO-2F system processes 26-group constants, whose resonance structure is represented by self-shielding factors or subgroup parameters. There is also provision for extending the number of groups of the system of cross sections toward higher energies.[87] In contrast with the previous versions, ARAMAKO-2F processes data on the anisotropy parameters of elastic scattering and prepares the cross sections required for solving the transport equation with an anisotropy of the scattering up to the P_5 approximation. The ARAMAKO-2F complex is used along with the ARAMAKO-G program complex to prepare cross sections for the interaction of neutron and γ radiations.[88]

The formats for representing the macroscopic and blocked microscopic cross sections in the computer storage media are described in detail in Refs. 86–88. Instructions on the use of the modules are also given there.

The algorithms for calculating the cross sections which are embodied in the ARAMAKO-2F program modules are described in Ref. 86.

The accuracy of the results calculated on a radiation field depends on the correspondence between the initial group cross sections and the most recent experimental data and on the quality of the algorithms for generating microscopic and blocked macroscopic cross sections from the initial group cross sections. The correct use of multigroup cross sections thus requires a good understanding of the approximations underlying the multigroup method and of the algorithms for preparing cross sections which are implemented in the existing program complexes. It is necessary to have an understanding of the errors in the calculated results which are related to these approximations. In order to evaluate the accuracy of calculated predictions, it is also necessary to have reliable information on the errors in the initial microscopic cross sections and on the errors which are actually introduced in a calculation of the macroscopic cross sections of the medium. The error in these macroscopic cross sections is determined to a large extent by how well two effects are taken into account: the resonant self-shielding of the cross sections in the real medium and the deviation of the actual intragroup spectrum from the standard spectrum which is adopted when the cross sections are averaged over the group intervals. This step has an important effect on the elastic moderation cross section $\sigma_{M(el)}$.

Correcting the moderation cross sections for the shape of the intragroup spectrum

The elastic-moderation cross sections $\sigma_{M(el)}$ are calculated by taking an average with the weight of a standard spectrum (the fission spectrum at energies above 1.4 MeV, a Fermi $1/E$ spectrum at $E \leqslant 1.4$ MeV, and a spectrum with a Maxwellian distribution in the thermal neutron region). This standard spectrum may be quite different from the neutron spectrum in the actual system for which the calculations are being carried out, especially in calculations on radiation shielding. Consequently, in the course of the calculations the elastic-moderation cross sections must be corrected by a factor b_g for the difference between the actual spectrum in the gth calculation zone and the standard spectrum. For the ith neutron energy group one can replace $\sigma_M^i(el)$ by $\sigma_M^i(el)b_g^i/\tilde{b}^i$, where the factor $b_g^i/\tilde{b}^i$ reflects the difference between the shape of the integral intragroup spectrum over the gth zone and the standard spectrum.

For nuclei which are not too light, we can write

$$b_g^i = \frac{\varphi_g(u_i - 2/3\xi^i)}{\varphi_g^i/\Delta u_i}, \tag{2.1}$$

where ξ^i is the average logarithmic energy loss, $\varphi_g(u)$ is the detailed shape of the integral spectrum, and

$$\varphi_g^i = \int_{\Delta u_i} \varphi_g(u)du. \tag{2.2}$$

The factor b_g^i thus depends on the particular isotope for which the correction is being made through ξ. For heavy isotopes, this is a weak dependence, since we have $\xi \ll \Delta u$ at $A > 100$.

A complexity of the calculation of the corrections b_g is that the multigroup calculation does not make it possible to estimate $\varphi_g(u)$, so that a calculation of b_g requires going beyond the scope of the multigroup calculation. A variety of methods are used to solve this problem in the existing program complexes. These methods may be put in two groups. The first group of methods for estimating $\varphi_g(u)$ uses the results of spectral calculations which are more detailed than the multigroup results (calculations of the spectra in the Grüling–Hertzel approximation and multigroup calculations). The second group of methods uses a description of the multigroup histogram of the integral spectrum of the zone by a more or less smooth curve (a piecewise-linear approximation). A disadvantage of the methods of the second group is that the convergence is slow in an iterative refinement of the multigroup histogram.

An advantage is that these methods can be used to estimate the intragroup spectral shape for the energy dependence of the harmonics of the radiation flux density. In other words, they can be used to solve a multigroup transport equation, as is necessary in calculations on neutron shielding.

Incorporating resonant self-shielding of the cross sections

This problem can be solved by making use of the concept of a dilution cross section σ_0 or by constructing a distribution function of macroscopic cross sections on the basis of specified subgroup parameters of the distribution functions of the cross sections of the isotopes which make up the medium.

The cross sections (σ_{0i}) for the dilution of each isotope ii which has resonances in its cross section by the other isotopes are found by an iterative procedure. In the kth iteration we have

$$\sigma_{0i} = \frac{1}{\rho_{ii}} \sum_{jj=1}^{I} \rho_{jj}\sigma_{t0,jj}^{(k-1)} - \sigma_{t0,ii}^{(k-1)}, \tag{2.3}$$

where

$$\sigma_{t0,jj}^{(k-1)} = \sigma_{el,jj}f_{el,jj}(\sigma_{0jj}^{k-1}) + \sigma_{c,jj}f_{c,jj}(\sigma_{0jj}^{(k-1)}) + \sigma_{f,jj}f_{f,jj}(\sigma_{0jj}^{k-1}) + \sigma_{in,jj}. \tag{2.4}$$

Here ρ is the concentration; σ_{t0}, σ_c, σ_f, σ_{el}, and σ_{in} are the total cross section, the capture cross section, the fission cross section, and the cross sections for elastic and inelastic scattering, respectively; $\sigma_{to} = \sigma_c + \sigma_f + \sigma_{el} + \sigma_{in}$ and f_c, f_f, f_{el}, and f_t are the resonant self-shielding factors for the capture, fission, elastic-scattering, and total cross sections for a given dilution cross section σ_0 (here and below, we are omitting the group index).

In the solution of the transport equation in the P_1 approximation it is recommended that the resonant self-shielding factors for the inelastic-scattering cross sections be taken to be equal to the self-shielding factors for the capture cross sections, rather than equal to unity as recommended in Ref. 89.

In calculations with the BNAB-78 cross sections one introduces the concept of self-shielding factors for the elastic-moderation cross sections for oxygen, defined by

$$f_{M(el)} = (\sigma_t f_t - \sigma_c f_c - \sigma_f f_f - \sigma_{in}f_c)/\sigma_{el}. \tag{2.5}$$

In a specification of the distribution function of the cross sections by means of subgroup parameters, self-shielding can also be taken into account with allowance for a scattering anisotropy.[90] The self-shielding factors of the cross sections are determined for temperatures in the range $300\ \text{K} < T < 2100$ K.

Use of group cross sections in solving the transport equation

The elastic-scattering anisotropy parameters can be used to carry out calculations in which the scattering anisotropy is taken into account in approximations up to P_5. In these calculations it should be kept in mind that in the case of a high flux gradient (and only in this case is it meaningful to consider a scattering anisotropy) the resonance structure of the energy dependence of the neutron flux density in a medium depends strongly on the angular coordinates. For this reason, the group-averaged total cross sections which appear in the equations of the method of spherical harmonics of the neutron flux are different for different harmonics of the neutron flux density. When the radiation transport equation is written in integrodifferential form, the anisotropy of the group-averaged total cross section gives rise to an integral term. The incorporation of this term is equivalent to the introduction of an anisotropic negative scattering cross section, which leaves the neutron within the given group after the scattering. This effect was noted in Refs. 86 and 90. Here we will simply call attention to the following characteristic features: (a) It is important to take the anisotropy of the total cross section into account, regardless of the accuracy of the incorporation of the scattering anisotropy. (b) The information on the structure of the total cross section which is embodied in the self-shielding factors is sufficient to take into account the anisotropy of the total cross section only in the transport approximation; subgroup constants would have to be used in order to take the scattering anisotropy into account in a higher-order approximation.[86] (c) The anisotropy of the total cross section can result in a poor convergence or even a divergence of the iterative process unless special measures are taken.[86] (d) The angular distributions of the neutron flux density which are reconstructed from six angular momenta (in the P_5 approximation) are not themselves positive definite at an arbitrary scattering angle, but the flux density of the radiation field which is integrated over the angular variables is always positive.

Errors in the group cross sections

For an estimate of the accuracy of calculations on the shielding or neutron-physics characteristics of a reactor, the system of cross sections which is used in the multigroup neutron calculations must be supplemented with the covariance matrix of the errors of these constants.

The BNAB-78 system of cross sections which is recommended for calculations also contains a covariance matrix of this sort. A method for estimating the covariance matrix of errors of the BNAB-78 cross sections is described in Refs. 91 and 92.

All the experimental data published through mid-1977 are incorporated in the BNAB-78 system. The most important data sets used in the estimate of the cross sections are taken into account in the estimates of the errors. For each such set q, independent errors of each experimental point n are estimated, and the total error of all the experimental points is estimated. In other words, the covariances of the errors of two different points in the qth set,

$$w^q_{nn'} = \overline{\delta\sigma_{qn}\delta\sigma_{qn'}} , \qquad (2.6)$$

due, for example, to an identical normalization error at all the measurement points, are assumed to be identical for different n and n'. In cases in which the values of $w_{nn'}$ for different pairs n and n' are greatly different, or where the maximum value of $w_{nn'}$ are comparable to $\sqrt{w_{nn}w_{n'n'}}$ and the publication contains enough information for a quantitative estimate of the covariances, the covariance matrix of the data set is constructed in more detail. The conversion of the preliminary estimates of the covariance matrices of the individual data sets into a covariance matrix of the group cross sections is carried out under the assumption that the process of estimating the group cross section σ_i reduces to taking an average of the data of the different sets:

$$\sigma_i = \sum_q \sum_n w_{qni}\sigma_{qni} . \qquad (2.7)$$

For experimental points within g, the weights w_{gni} are usually determined by the errors of the measurements (roughly speaking, w_{gni} is inversely proportional to the square of the error of σ_{gni}). An estimate of $\sigma(E)$ is frequently found by drawing curves through experimental points "by hand." In such cases, the weights w_{gni} may be determined very crudely. An arbitrariness in the determination of the weights w_{gni} of course reduces the accuracy of the estimate of the errors. Experience shows, however, that the uncertainty in the estimate of the standard deviations which stems from this arbitrariness does not exceed 30%.

Just how realistic the resulting estimates of the errors and correlations of the group cross sections are has been checked by comparing the results of different estimates based for the most part on the same experimental material. Comparisons have been carried out with data furnished by the Nuclear Data Center and the data in the ENDF/B-IV, UKNDL, KEDAK, and JENDL libraries.

Discrepancies which exceed our estimates of the errors are extremely rare. A comparison has also been made of the results of estimates of the covariance matrices found by various investigators.[92] The error covariance matrix which we have found is intended for use in estimating the errors of calculated predictions of the neutron-physics characteristics of fast-neutron reactors and of shielding. At present, covariance matrices have been constructed for the fission cross sections of ^{235}U, ^{238}U, ^{239}Pu, ^{240}Pu, and ^{241}Pu; for the average number of fission neutrons from the nuclei of these isotopes; for the radiative-capture cross sections of ^{235}U, ^{238}U, ^{239}Pu, ^{240}Pu, ^{241}Pu, Na, Fe, Cr, and Ni; for the inelastic-scattering cross sections of ^{235}U, ^{238}U, ^{239}Pu, Na, Fe, Cr, and Ni; for the elastic cross sections of Na, Fe, O, and C; for the total cross sections of Fe, Cr, Ni, Na, O, and C; and for the average cosine $\langle\mu_{tl}\rangle$ of the elastic-scattering angle of these isotopes.

2. Programs for calculating radiation fields, calculation errors, and sensitivity of results to the input parameters

Practical applications of the mathematical apparatus of sensitivity analysis and estimates of computational errors require suitable programs for calculating the radiation fields in shielding, their functionals, and the sensitivities of these functionals to the input parameters of the calculations; for storage in files of all the computational data, the results of benchmark experiments, the covariance matrix of their errors, the covariance matrix of the errors of the group cross sections, etc.; and for performing the matrix-algebra operations required for solving the particular problem on the matrices and vectors in the files. The best-known non-Soviet system of this type is the FORSS system.[57] In the Soviet Union, such problems are solved by the INDÉKS system, whose name is an acronym formed from the Russian words for one of the most general and laborious tasks of sensitivity analysis: "correcting neutron data on the basis of an analysis of the results of macroscopic experiments." The word "system" here means a set of programs, developed independently for various purposes, which are used on the same powerful computer (the BÉSM-6) along with a common system for generating cross sections, which communicate with each other through computer files, and which together offer functional completeness; i.e., they provide the tools for solving that class of sensitivity-analysis problems which is of interest and for estimating the errors in the calculated results.

In the current version of the INDÉKS system, the cross sections are generated from three permanent libraries of cross sections and the software which services these libraries. Together, the libraries and the software form the SOKRATOR system (an acronym formed from the Russian words for "system for providing constants for calculations on atomic reactors and radiation shielding").[93,94] The basic library is the FOND library (an acronym formed from the Russian words for "files of evaluated neutron data"), written in text form in the ENDF/B format.[95] The current version of this library corresponds primarily to the data which were used to generate the BNAB-MIKRO system of 28-group cross sections.[96] The data from the FOND library and, where necessary, other libraries of estimated data are processed into group cross sections by the GRUKON applications software packet (GRUKON is an acronym from the Russian words for "calculation of group constants").[94,97] The combination of FOND with GRUKON makes it possible to update and add to the contents of a second constant library: the library of group cross sections of the ARAMAKO system,[86,88] developed for providing cross sections for multigroup calculations on radiation fields in reactors and shielding. In the current version of the ARAMAKO system, it is possible in practice to carry out 26-, 28-, or 49-group neutron calculations and calculations of the γ fields in a 15-group approximation. The standard constant base is the BNAB-78 system of cross sections,[96] partitioned where necessary in more detail at high neutron energies.[86,96] The data from the group libraries are processed into the macroscopic cross sections required for calculations of the radiation fields, and the cross sections required for calculating the functionals of these fields are generated, rapidly by the ARAMAKO system, included for convenience in controlling the specification in the OKS applications software packet (this name is an acronym formed from the Russian words for "combined constant system").[98]

The third cross section library is the LUND ("library of uncertainties of nuclear data") library.[29,99] This library consists of the covariance matrices of the errors in the group cross sections of the BNAB-MIKRO system, from which the BNAB-78 system, used here, differs only in certain of the ^{238}U cross sections, corrected on the basis of experimental data from fast critical assemblies.[96] A description of the evaluation of these covariance matrices and the corresponding numerical data have been published as a handbook.[92] In addition to the covariance matrix of errors in the group cross sections of the basic library, the LUND library can store the covariance matrices of the errors of various versions of corrected cross sections.

A second large group of programs in the INDÉKS system constitute the programs for calculating the radiation fields, their functionals, and the relative sensitivities. The INDÉKS system was developed for analyzing the errors in the results of calculations on radiation shielding and on the core of nuclear reactors, in particular, fast-neutron reactors. The basic programs used in calculating the fields in radiation shielding are the programs of the ROZ series (the name is an acronym from the Russian words for "calculation of one-dimensional shielding").[100-103] These programs calculate the neutron and γ fields and their functionals for shields of one-dimensional geometry. The authors of the present book have used the ROZ-11 program[103] for calculations.

The ROZ-11 program carries out a numerical solution in plane geometry with azimuthal symmetry or in spherical geometry of the following system of basic transport equations, with boundary conditions corresponding to specular reflection or to radiation incident from the exterior:

$$\overline{\Omega\nabla}\varphi^{i}(r,\mu) + \Sigma_{t}^{i}(r)\varphi^{i}(r,\mu) - \sum_{j=1}^{i} \int_{2\pi}\int_{-1}^{1} \Sigma_{S}^{j\rightarrow i}(r,\mu_{S})\varphi^{j}(r,\mu')d\varphi' \, d\mu'$$

$$= S^{i}(r,\mu) + \begin{cases} \dfrac{S_{a}^{i}\delta(1-\mu)\delta(r-a)}{2\pi} - \dfrac{S_{A}^{i}\delta(1+\mu)\delta(r-A)}{2\pi}, \text{ for plane geometry;} \\ \dfrac{S_{a}^{i}\delta(1-\mu)\delta(r)}{2\pi4\pi r^{2}}, \text{ for spherical geometry;} \end{cases}$$

$$(2.8)$$

$$r\in[a,A] \, ;$$

$$\varphi^{i}(a,\mu) = \varphi^{i}(a, -\mu) \text{ or } \varphi^{i}(a,\mu) = \varphi_{a}^{i}(\mu) \text{ for } \mu > 0 \, ;$$

$$\varphi^{i}(A,\mu) = \varphi^{i}(A, -\mu) \text{ or } \varphi^{i}(A,\mu) = \varphi_{A}^{i}(\mu) \text{ for } \mu < 0 \, ;$$

$$i = \overline{1,I} \, .$$

Here $\varphi^{i}(r,\mu)$ is the spatial energy-angular flux density of the particles of the ith energy group; $\Sigma_{t}^{i}(r)$ is the total macroscopic cross section for the interaction of the particles of the ith energy group; $\Sigma_{S}^{j\rightarrow i}(r,\mu_{S})$ is the doubly differen-

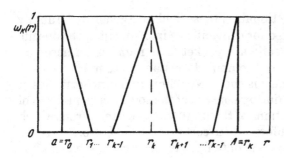

Figure 5. The coordinate functions $\omega_k(r)$ used in the ROZ-11 program to describe the spatial dependence of the solution of the transport equation, of the source specification function, and of the detector response function.

tial cross section for the scattering of particles through an angle $\theta = \arccos \mu_S$ from group j into group i; $S^i(r,\mu)$ is the spatial energy-angular density of the sources of the particles of group i; S^i_α and S^i_A are the numbers of particles of the ith energy group of the source which are incident in a unidirectional fashion on the shielding per unit time; $\varphi^i_\alpha(\mu)$, $\varphi^i_A(\mu)$ are the energy-angular flux densities of the particles incident on the shielding; and I is the number of groups in the calculation.

The angular dependence of the quantities $S^i(r,\mu)$, $\varphi^i_\alpha(\mu)$, $\varphi^i_A(\mu)$ and $\Sigma^{j-i}_S(r,\mu_S)$ is described by a finite sum of Legendre polynomials. In terms of the spatial variable, $S^i(r,\mu)$ is a piecewise-linear function with possible discontinuities at the boundaries of the geometric zones, which may differ in density and composition. The cross sections Σ^i_t and Σ^{j-i}_S have constant values in geometric zones $m = 1,..., M$. If unidirectional sources S^i_α and S^i_A are given at $r = \alpha$ or $r = A$, the unscattered radiation is calculated from analytic formulas, while the spatial dependence of the sources of the first collisions is specified by a piecewise-exponential or piecewise-linear interpolation with possible discontinuties at the boundaries of the geometric zones. The calculation of the unscattered neutron radiation and of the sources of the first collisions can be carried out through the use of both blocked neutron-matter interaction cross sections or unblocked cross sections. The scattered neutron radiation is calculated through the use of blocked cross sections. There is also provision for numerical solution of the adjoint problem.

The ROZ-11 program implements a variational-difference method with coordinate functions[104,105]

$$\varphi^i_{kn}(r,\mu) = \frac{\omega_k(r)P_n(\mu)}{\sqrt{\Sigma^i_t(r)}a_{kn}} \qquad k = 0,...,K, \quad n = 0,...,N^i, \qquad (2.9)$$

where K is the number of steps in the spatial grid, $N^i + 1$ is the number of Legendre polynomials used in describing the angular dependence of the solution of the transport equation for the ith group, $\Sigma^i_t(r)$ is the total macroscopic cross section for the interaction of the radiation of the ith group with matter, a_{kn} is a normalization factor, the P_n are the Legendre polynomials, and the ω_k are coordinate functions for describing the spatial dependence (Fig. 5). The solution $\varphi_i(r,\mu)$ is written in the form

$$\varphi^i(r,\mu) = \sum_{k=0}^{K} \sum_{n=0}^{N^i} \alpha^i_{kn} \varphi^i_{kn}(r,\mu). \qquad (2.10)$$

The ROZ-11 program has a module structure and a memory allotment which is dynamic in terms of FORTRAN. It is written in FORTRAN IV for the BÉSM-6 computer. The dynamic allotment of memory makes it possible to vary the spatial and angular approximations over broad ranges. The following relation holds approximately:

$$(K+1)\{3+(N^i+1)/2\,[(N^i+1)/2+8)\,]\} \leqslant N_w - 2000\,, \qquad (2.11)$$

where N_w is the number of words of computer memory which are allotted to a file of the program (this number is specified in the control program). For the BÉSM-6 computer, for example, we would have $N_w \leqslant 16\,000$ with a memory expansion to 32 sheets. The maximum number of groups in a calculation is determined by the external memory. The maximum number of geometric zones is determined by the following condition: The total number of nodes over all the zones (m) must not exceed the maximum possible number of nodes give by expression (2.11). Programs other than those of the ARAMAKO-2F and ARAMAKO-G systems could be used; the only requirement which must be met is that they have an output file in the FMAC5A format of the ARAMAKO-2F system.

Relative sensitivities are calculated by the ZAKAT program (the name is an acronym formed from the Russian words for "shielding channel theories").[56] The program for calculating the relative sensitivity uses equations from a generalized perturbation theory for the linear functionals of the radiation field in shielding.[31,39] The ZAKAT program[56] was written for calculating the relative sensitivity of functionals of the radiation field to input parameters. It implements energy and spatial channel theories in a plane geometry with azimuthal symmetry and in a spherical geometry. The ZAKAT program calculates the following quantities.

(1) The energy dependence of the relative sensitivity of functional R to the absorption cross section of the κth isotope, $\Sigma_c^{\kappa i}(r)$, in spatial zone V_m:

$$p_{Rc}^{\kappa}(i,V_m) = -\frac{1}{R}\int_\Omega\int_{V_m} \varphi^{\,i}(r,\mu)\Sigma_c^{\kappa i}(r)\varphi^{\,i*}(r,\mu)d\Omega\,dV\,, \qquad (2.12)$$

where $\varphi^{i*}(r,\mu)$ is the importance function of the particles of group i.

(2) The energy dependence of the relative sensitivity of functional R to the harmonics (and to their sum) of the cross section for elastic (inelastic) scattering by the κth isotope, $\Sigma_{\mathrm{el(in)}l}^{\kappa i\to j}$, $\Sigma_{\mathrm{el(in)}l}^{\kappa i}$, and $\Sigma_{\mathrm{el(in)}}^{\kappa i}$ in spatial zone V_m:

$$p_{\mathrm{Rel(in)}}^{\kappa}(i\to j,l,V_m)$$

$$=\begin{cases} -\dfrac{1}{R}\displaystyle\int_\Omega\int_{V_m}\varphi^{\,i}(r,\mu)\Sigma_{\mathrm{el(in)}0}^{\kappa i\to j}(r)\varphi^{\,i*}(r,\mu)d\Omega\,dV \\[2mm] +\dfrac{1}{4\pi R}\displaystyle\int_{V_m}\varphi^{\,i}_0(r)\Sigma_{\mathrm{el(in)}0}^{\kappa i\to j}(r)\varphi^{\,i*}_0(r)dV, \quad l=0; \\[2mm] \dfrac{2l+1}{4\pi R}\displaystyle\int_{V_m}\varphi^{\,i}_l(r)\Sigma_{\mathrm{el(in)}l}^{\kappa i\to j}(r)\varphi^{\,i*}_l(r)dV, \quad l\neq0; \end{cases} \qquad (2.13)$$

$$p_{\mathrm{Rel(in)}}^{\kappa}(i,l,V_m) = \sum_{j=i}^{I} p_{\mathrm{Rel(in)}}^{\kappa}(i\to j,l,V_m);$$

$$p_{\mathrm{Rel(in)}}^{\kappa}(i,V_m) = \sum_{l=0}^{L_{\max}} p_{\mathrm{Rel(in)}}(i,l,V_m)\,.$$

(3) The energy dependence of the relative sensitivity of functional R to the total cross section of the κth isotope, $\Sigma_t^{\kappa i}(r)$, in spatial zone V_m:

$$p_{Rt}^\kappa(i, V_m) = p_{Rc}^\kappa(i, V_m) + p_{Rel}^\kappa(i, V_m) + p_{Rin}^\kappa(i, V_m). \qquad (2.14)$$

(4) The energy dependence of the relative sensitivity of the functional R to the source specification function and the detector response function, lumped at the shielding boundaries:

$$\left.\begin{array}{l}
p_{RS}(i) = \dfrac{s(a(A))}{R} \displaystyle\int_\Omega S_{a(A)}^i \varphi^{i*}(r = a(A), \mu) d\mu; \\[4mm]
p_{Rq}(i) = \dfrac{s(a(A))}{R} \displaystyle\int_\Omega q_{a(A)}^i \varphi^i(r = a(A), \mu) d\mu,
\end{array}\right\} \qquad (2.15)$$

where $s(r) = 1$ in plane geometry and $s(r) = 4\pi r^2$.

(5) The spatial-energy dependence of the response current:

$$|\mathbf{R}(r, i)| = \frac{s(r)}{R} \left| \int_\Omega \mu \varphi^i(r, \mu) \varphi^{i*}(r, \mu) d\Omega \right|. \qquad (2.16)$$

We wish to stress that these expressions are valid if cross sections of a given type—i.e., either blocked or unblocked—are used in the calculations. If cross sections of both types are used in the calculations, expressions (2.12)–(2.16) must be modified. This is the case in the ZAKAT program. In this case it is necessary to solve two adjoint problems. Since no fundamental difficulties arise in a generalization of expressions (2.12)–(2.16) or in the derivation of equations for the two adjoint problems, we will not present them here.

The ZAKAT program has a modular structure and is written, like the ROZ-11 program, in FORTRAN IV for the BÉSM-6. It has a dynamic allottment of memory in terms of FORTRAN, and there is provision for direct access in operation with external devices.

A certain part of the functional capabilities of the INDÉKS system consists of programs which are specialized for multigroup calculations for reactors. Among them are the NULGEO program for calculations in the B_1 approximation, the KRAB program for calculations on one-dimensional reactors in the P_1 and S_n approximations, the TVK-2D program for diffusion calculations on reactors in the XY geometry and the RZ geometry, and programs for calculations on reactors and shielding by the Monte Carlo method in the real three-dimensional geometry.[106] Each of these programs is a packet consisting of subprograms for calculating neutron fields, for calculating various functionals of these fields of practical importance, and for calculating the sensitivities of the basic functionals to the macroscopic constants.

The third part of the INDÉKS system is formed by a system of programs for analyzing relative sensitivities and for estimating the errors in functionals of the radiation fields: the CORE packet. Its structure is shown in Fig. 6. Most of the data used by the programs of this packet are stored in three computer libraries: the LUND error library, the LSENS library of relative sensitivities, and the LEMEX library of evaluated data from macroscopic benchmark experiments. Each library has its own software support for storing new information, for adding to or revising information which has already been stored, for generating information on the contents of the library, and for pro-

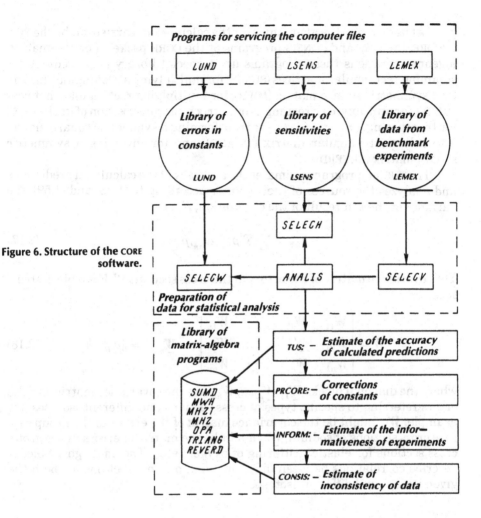

Figure 6. Structure of the CORE software.

grammed extraction of the necessary information. These procedures are used by the central program of the system, the ANALIS, which is the software for preparing data for subsequent statistical analysis.

The LSENS library is filled by software; the results of calculations carried out by means of the sensitivity-calculation programs are stored in this library by means of special interface programs. The LEMEX library is filled manually. The LEMEX stores the names of the experiments; the names of the functionals measured in the experiments; the experimental results, including all the corrections required for referring these results to the experimental model for which the calculations are being carried out (corrections of type A) and a covariance matrix of the errors in the experimental data (with allowance for the errors in the corrections of type A); and the calculated results, including all the corrections required for referring them to the experimental model for which calculations are being carried out (corrections of type B) and a covariance matrix of the errors in the calculated results (with allowance for the errors in the corrections of type B).

Let us examine the algorithms for the calculations carried out by the TUS, PRCORE, INFORM, and CONSIS programs of the CORE packet. For the matrix-algebra operations these programs use a special library of procedures for summing rectangular matrices (SUMD), for multiplying a rectangular matrix by a symmetric square matrix (MWH), for multiplying rectangular matrices (MHZT), for multiplying a rectangular matrix by a transposition of itself (MHZ), for transposing a matrix (OPA), for transforming a symmetric square matrix into a lower triangular matrix (TRIANG), and for inverting a symmetric square matrix (REVERD).

The TUS subprogram estimates the accuracy of the calculated predictions and analyzes the sources of their errors. According to (1.48) and (1.59), the variance of the kth result of the calculation, R_k, is

$$u_{kk} = \sum_n \sum_{n'} p_{R_k n} w_{nn'} p_{R_k n'}. \tag{2.17}$$

The covariance matrix of the errors in the cross sections $\underline{W}$, has a block structure:

$$\underline{W} = \begin{vmatrix} \underline{W}_{11} & \underline{W}_{12} & \cdots & \underline{W}_{1n} \\ \underline{W}_{21} & \underline{W}_{22} & \cdots & \underline{W}_{2n} \\ \underline{W}_{n1} & \underline{W}_{n2} & \cdots & \underline{W}_{nn} \end{vmatrix}, \quad \underline{W}_{mm'} = \| w_{ll}^{mm'} \|, \tag{2.18}$$

where the diagonal blocks $\underline{W}_{11}, \underline{W}_{22},..., \underline{W}_{nn}$ are the covariance matrices of the errors referring to specific types of cross sections and different isotopes (for example, $\underline{W}_{11}$ might be the covariance matrix of the errors in the group capture cross sections of sodium, $\underline{W}_{22}$ might be the matrix of errors in the group cross sections for elastic scattering of oxygen, etc.). The off-diagonal blocks describe correlations between different types of cross sections for both the given isotope and different isotopes.

Example
Let us assume that we are interested in the error in the calculated prediction of the activation of sodium in the loop of a fast reactor, and we wish to calculate the cross section component of this error, which stems from the uncertainties in the capture, elastic, inelastic, and total cross sections of the shielding materials: isotopes of steel (Fe, Cr, and Ni). The matrix of errors W can then be represented as follows. The diagonal blocks of the matrix W are the covariance matrices of the errors in the capture, elastic, inelastic, and total cross sections of Fe, Cr, and Ni. The off-diagonal blocks W_{ml} ($m \neq l$) are nearly all zero. The nonzero blocks are those which reflect correlations between errors in the total cross section and in the elastic cross section separately for each of the isotopes of Fe, Cr, and Ni by virtue of the relation

$$\delta\sigma_{t0} = \delta\sigma_c + \delta\sigma_f + \delta\sigma_{el} + \delta\sigma_{in} \approx \delta\sigma_{el}.$$

Table IV shows that the LUND library contains covariance matrices of the errors in the constants of Fe, Cr, and Ni and blocks of correlations between the errors in constants of different types.

Table IV. Blocks of covariance matrices of the errors in the cross sections for isotopes of Fe, Cr, and Ni in the LUND library.

Cross section and isotope		1	2	3	4	5	6	7	8	9	10	11	12
σ_{t0} Fe	1	+		+									
σ_c Fe	2		+				+				+		
σ_{el} Fe	3	+		+									
σ_{in} Fe	4				+				+				+
σ_{t0} Cr	5												
σ_c Cr	6		+				+				+		
σ_{el} Cr	7												
σ_{in} Cr	8				+				+				+
σ_{t0} Ni	9												
σ_c Ni	10		+				+				+		
σ_{el} Ni	11												
σ_{in} Ni	12				+				+				+

Note. The plus sign means that data are present; an empty cell means that there are no data.

The contributions from the various types of cross sections to the total error in the calculated prediction are evaluated in the TUS program. The variance of the kth result of the calculation due to the uncertainties in the cross sections of the mth type is

$$w_{kk}^{mm} = \sum_l \sum_{l'} P_{R_k l} w_{ll'}^{mm} P_{R_k l'} . \tag{2.19}$$

Under the assumption of weak correlations between cross sections of different types, the total variance of the calculated prediction, u_{kk}, is equal to the sum of the contributions u_{kk}^m (this assumption holds for most types of cross sections).

The PRCORE subprogram implements algorithms for correcting the cross sections on the basis of the results of benchmark experiments. New, refined estimates of the cross sections are found from Eq. (1.57). The covariance matrix of refined values of the cross sections is calculated in accordance with Eq. (1.58).

The INFORM subprogram evaluates the informativeness of benchmark experiments. The informativeness of the $(J+1)$-th experiment is estimated as the decrease in the variance of the calculated prediction of the functional R_k of interest:

$$I_r(R_k) = u_{kk}(J) - u_{kk}(J+1)$$
$$= H_k W H_{J+1}^T (v + H_{J+1} W H_{J+1}^T) H_{J+1} W H_k^T , \tag{2.20}$$

where H_k is the column vector of relative sensitivities of the functional R_k, H_{J+1} is the column vector of relative sensitivities of the $(J+1)$-th experiment, v is its variance, and W is the covariance matrix of the cross sections corrected on the basis of the results of J measurements (not including the measurement whose informativeness is being evaluated). Several values of the relative informativeness I_r are calculated for several values of the variance of the experiment in question, v. Analysis of I_r and of its dependence on v can show whether it is advisable to formulate the given experiment from the standpoint of the optimum relation between the informativeness I_r and the required experimental accuracy v (the informativeness of an experiment may turn out to be too low to justify the cost of carrying out the experiment).

The CONSIS subprogram estimates the statistical inconsistency of the experimental and calculated results being analyzed. Two criteria are used for this purpose. The first, integral, criterion is the minimum value (S^2_{min}) found for the quadratic form in Eq. (1.54), which, as a function of random quantities, is itself a random quantity. If the linear hypothesis is valid and if there are no systematic errors in the calculated and experimental results, the distribution of this minimum value of the quadratic form obeys a χ^2 distribution with J degrees of freedom, where J is the number of functionals considered. If the minimum value S^w_{min} of functional (1.54) differs from the expected value (J) by more than one standard deviation, $(\sqrt{2J})$, one would suspect that there are unidentified systematic errors in the data being analyzed.

The second, differential, criterion for inconsistency of the data is a more powerful tool. Let us consider the quadratic form which is constructed from the differences between the results of macroscopic experiments, R^0, and the results of calculations from the corrected cross sections $\underline{R} + \hat{H}\Delta\underline{Y}$:

$$\zeta^2 = (\underline{R}^0 - \underline{R} - \hat{H}\Delta\underline{Y})^T(\hat{\underline{M}} + \hat{H}\hat{\underline{W}}\hat{H}^T)^{-1}(\underline{R}^0 - \underline{R} - \hat{H}\Delta\underline{Y}). \qquad (2.21)$$

We decompose the matrix $(\hat{\underline{M}} + \hat{H}\hat{\underline{W}}\hat{H}^T)^{-1}$ into the product of two triangular matrices:

$$(\hat{\underline{M}} + \hat{H}\hat{\underline{W}}\hat{H}^T)^{-1} = \hat{\underline{A}}^T\hat{\underline{A}}. \qquad (2.22)$$

We can then write

$$\zeta^2 = [\hat{\underline{A}}(\underline{R} - \underline{R}^0 - \hat{H}\Delta\underline{Y})]^T\hat{\underline{A}}(\underline{R} - \underline{R}^0 - \hat{H}\Delta\underline{Y}) = \underline{Q}^T\underline{Q}.$$

In the absence of systematic errors in the results of the estimate of the macroscopic experiments, the components of the vector $\underline{Q} = \{q_j\}$, $j = 1,2,...,J$ have a normal distribution with a zero mean and a unit variance. The correspondence between the observed distribution of the components q_j and the expected distribution is tested on the basis of the Kolmogorov criterion.[77] If substantial deviations—"substantial" in the sense of the Kolmogorov criterion—of the distribution of the components q_j from a normal distribution is found, a detailed analysis of the specific nature of the deviations usually makes it possible to identify those functionals for which the results of the measurements or calculations in all probability contain unidentified systematic errors.

Considerable experience has now been acquired in analyzing the sensitivities of shielding and reactor characteristics to the interaction cross sections; we might cite Refs. 29 and 106–108 as examples. This experience has shown that the INDÉKS system is a quite versatile and convenient tool for sensitivity analysis.

3. Models for benchmark calculations

Studies of the sensitivity of calculated results to the initial parameters of the problem are used widely in various fields of research on reactor shielding. We will discuss here only the most typical problems for which calculations have been carried out by sensitivity analysis.[14,29,30,84,109–127]

Figure 7. Geometry of the calculation models. Notation for problems 10–12: 1, concrete shielding; 2, wall; 3, austenite cladding; 4, second water layer; 5, jacket (heat shield); 6, first water layer; 7, reflector; 8, core; 9, lateral shielding; 10, sodium tank; 11, heat exchanger; 12, sodium of second loop. The dimensions are in centimeters.

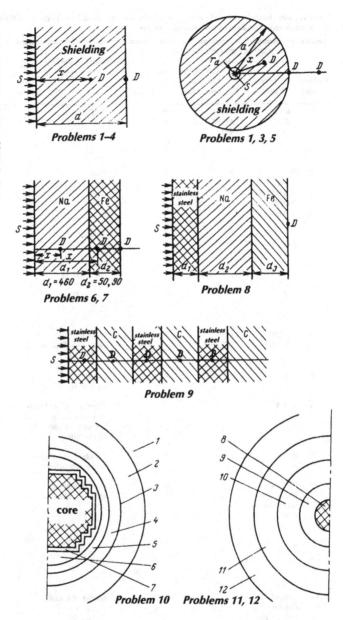

Table V shows the characteristics of the basic computational studies which have been carried out on the analysis of the sensitivity of the results to the radiation-matter interaction cross sections, to the source specification function, to the detector response function, and to the layout of the shield. Figure 7 shows the geometric configurations for these problems. It is convenient to classify these calculations on the basis of the shield layout: homogeneous shields (problems 1–5), heterogeneous shields (problems 6–9), and mock-ups of real shields for thermal- and fast-neutron reactors (problems 10–12).

Table V. Basic characteristics of the models used in the benchmark calculations of the sensitivity of the functionals of the neutron-radiation field to the input parameters.

Problem (model)	Characteristics of neutron source		Characteristics of shield		Functional calculated
	Energy distribution	Geometry and angular distribution	Geometry	Material (thickness, m)	
1	$f + 1/E$, INC	PI;Pt; PC;PC + 3, PU	P S	Fe(0.5); Na(4.6)	$P;P_{equ};\varphi;$ $\varphi_F;\varphi_{ind};$ $\varphi_{1/v};\varphi(x,E)$
2	f; f for E_0 < 0.9 MeV; 0 for E_0 > 0.9 MeV	PI	P	Na(5)	φ
3	$E_0 = 2$ MeV; $E_0 = 14$ MeV	PI; Pt	P; S	Fe(0.22;1)	$\varphi(E)$
4	f	PI	P	Fe(20.3;50.8; 76.2;101.6)	$\varphi(E)$
5	^{252}Cf	Pt	S	Fe(0.2;1)	$\varphi;\varphi_F;P_{equ}$
6	$f + 1/E$; INC;FTR	PI;PC; PC3;PU	P	Na(4.6); Na(4.6) + Fe(0.9); + Fe(0.5)	$P;P_{equ};\varphi;$ $\varphi_F;\varphi_{ind};$ $\varphi_{1/v};\varphi(x,E)$
7	FTR	PI	P	Na(4.6) + Fe(0.9)	P
8	TSR-II	PI	P	Combination of layers of stainless steel, sodium, and iron (Table VII)	Bonner detector
9	B-2	PU	P	Combination of steel and graphite layers	$RD;P_{equ};HE$
10	Core of pressurized water reactor, PWR		C	Model of actual shield (Table VII.)	$RD;P_{equ};$ HE
11	Core of fast-neutron reactor, FBR		S	Model of actual shield (Table VII)	AS,RD
12	Core of fast-neutron reactor, FN		S	Model of actual shield (Tables VII; VIII)	VDM

[a] Parameter for which sensitivity function was calculated.
[b] The system of constants of the SABINE program was used.

Table V. (continued)

	Nature of calculation software			Characteristics of sensitivity-analysis software			
Calculation method	Program	Approximations	System of constants	Parameter[a]	Calculation method	Program	Ref.
VDM	ROZ-11	$P_9P_5;P_9P_3;$ $P_9P_0;P_9P_1$	ARAMAKO-2F	$\Sigma_t, \Sigma_a,$ $\Sigma_{el}, \Sigma_{in},$ Characteristics of source and detector	DS, GPT	ZAKAT	[109, 110]
DO	ANISN	—	ENDF/B-IV	Σ_t, Characteristic of source	GPT	SWANLAKE	[84,111]
MM	—	—	KFKGGA	Σ_t, Characteristic of source	DS	—	[14]
DO	ANISN	$S_4P_3;S_8P_3;$ $S_{16}P_3;S_{16}P_2$	EURLIB UKNDL	Σ_t	DS, GPT	SWANLAKE	[112]
DO VDM	ANISN; ROZ-11	$S_{12}P_3;$ P_9P_5	EURLIB ARAMAKO-2F	$\Sigma_t, \Sigma_a, \Sigma_s$	DS GPT	SWANLAKE ZAKAT	[113,114, Present study]
VDM	ROZ-11	P_9P_5	ARAMAKO-2F	$\Sigma_t, \Sigma_a, \Sigma_{el},$ Σ_{in}, Characteristics of source and detector	DS GPT	ZAKAT	[115]
DO	ANISN	—	ENDF/B-III	Σ_t	DS GPT	SWANLAKE	[116, 117]
DO	ANISN DOT-III	—	ENDF/B-IV	Σ_t	GPT	SWANLAKE	[118]
VDM	ROZ-11 ROZ-5	P_9P_5	ARAMAKO-2F	$\Sigma_t, \Sigma_a,$ Σ_{el}, Σ_{in}	GPT DS	ZAKAT	[30, 119]
DO	ANISN	$S_8P_0;S_8P_1;$ $S_8P_2;S_8P_3$	EURLIB	Σ_t	GPT, DS	SWANLAKE	[120, 121]
DO	ANISN	$S_4P_1;S_4P_3$	DLC-2D	Σ_t	DS, GPT	SWANLAKE	[122]
DE DO	SABINE-3 DTF-IV	S_4	SABINE[b] MULCOS	Σ_t, Σ_a for iron	DS	—	[123]
DO	ANISN	$S_6P_0;S_6P_1;$ $S_6P_2;S_6P_3$	EURLIB-IV	Σ_t	GPT	SWANLAKE	[124]
DO	ANISN	$S_3P_0;S_3P_1;$ S_3P_2	—	Σ_t	GPT	SWANLAKE	[125]
DO	ANISN	$S_3P_0;S_1P_2$	—	Σ_t	GPT	SWANLAKE	[125]
DO	ANISN	S_4P_1	EURLIB-III	Σ_t	GPT	SWANLAKE	[126]
DO	ANISN	$S_4P_1;S_4P_3$	EURLIB	$\Sigma_t, \Sigma_a,$ Σ_{in}	GPT	SWANLAKE	[127]
VDM	ROZ-11	P_9P_5	ARAMAKO-2F	Σ_t	GPT	ZAKAT	[29]

Each problem is characterized by its particular geometry, the angular and energy distributions of the radiation from the source, the characteristics of the shield (its geometry, material, layout, and thickness), the characteristics of the radiation field under study (the functionals which are calculated), inside the shield or beyond it, the calculation software, and the characteristics of the sensitivity-analysis software.

Let us examine in more detail the characteristics of the models used in the benchmark calculations reflected in Table V.

Characteristics of the source

Energy distribution of the source neutrons

This distribution spans a rather broad range of neutron energies: (a) a neutron fission spectrum f; (b) a spectrum of neutrons from the spontaneous fission of a ^{252}Cf source; (c) the neutrons from monoenergetic sources with an energy E_0 of 2 or 14 MeV; (d) the spectrum of leakage neutrons from the core of a reactor (in Table V, INC is an intermediate-neutron converter[128]; FTR is a fast test reactor[116,117]; TSR is a tower shielding reactor[118]; B-2 is an installation for studying shielding in the B-2 beam of the BR-10 fast-neutron reactor[119]); and (e) the spectrum $f + 1/E$ (a fission spectrum at $E \geqslant 0.8$ MeV and a $1/E$ spectrum at $E < 0.8$ MeV), which is frequently used for an averaging of multigroup cross sections. The source neutron spectra shown here are shown in a 26-energy-group representation in Table VI.

In benchmark problems modeling the actual design of the shielding in pressurized water reactors (PWR), fast breeder reactors (FBR), and fast-neutron reactors, the core of the reactor, the neutron source, is part of the calculation physical zone. It may be assumed that the spectra of the neutrons emitted from the core of a reactor are close to an $f + 1/E$ spectrum for the PWR and close to an FTR (or INC) spectrum for FBR and fast-neutron reactors.

Geometry and angular distribution of the source neutrons

The characteristics of the radiation field in plane shields and at the exit from barriers were calculated in Refs. 109, 110, and 115 for four angular distributions of the source neutrons, spanning the entire range of angular distributions of plane sources which are possible in practice: a plane isotropic (PI) source $f(\theta) = 1/2\pi$; a plane cosinusoidal (PC) source $f(\theta) = \cos\theta/\pi$; a plane cosine-cubed (PC3) source $f(\theta) = 2\cos^3\theta/\pi$; and a plane unidirectional (PU) source $f(\theta) = \delta(\theta - 0)$, where $\theta \in [0°, 90°]$ is the angle measured from the normal to the surface of the barrier.

The studies which have been carried out in spherical geometry have generally used an isotropic point source (Pt) $f(\theta) = 1/4\pi$.

Table VI. Energy distributions of the neutrons from various sources in 26 energy groups, expressed in units of neutrons[a] per square centimeter per second and neutrons[a] per second for plane and point sources, respectively.

Group	Energy of upper boundary, MeV	$f+1/E$	INC	FTR	f	B-2	^{252}Cf
1	10.5	9.11 − 4[b]	5.01 − 7	3.35 − 6	1.6 − 2	1.3 − 3	2.1 − 2
2	6.5	5.03 − 3	7.76 − 7	1.77 − 5	8.8 − 2	4.76 − 3	8.7 − 2
3	4.0	1.04 − 2	3.35 − 6	6.49 − 5	1.84 − 1	1.19 − 2	1.9 − 1
4	2.5	3.30 − 2	4.73 − 5	2.26 − 4	2.70 − 1	5.00 − 2	2.75 − 1
5	1.4	3.19 − 2	1.09 − 3	6.10 − 4	2.02 − 1	1.00 − 2	2.15 − 1
6	0.8	3.95 − 2	4.75 − 2	9.51 − 3	1.41 − 3	1.19 − 1	1.6 − 1
7	0.4	3.95 − 2	1.25 − 1	5.25 − 2	6.1 − 2	1.55 − 1	6.4 − 2
8	0.2	3.95 − 2	2.32 − 1	9.16 − 2	2.4 − 2	6.07 − 2	—
9	0.1	4.36 − 2	1.08 − 1	1.10 − 1	1.0 − 2	3.81 − 2	—
10	4.65 − 2	4.40 − 2	4.12 − 1	1.19 − 1	3.0 − 3	2.98 − 2	—
11	2.15 − 2	4.36 − 2	3.23 − 2	1.01 − 1	1.0 − 3	3.21 − 2	—
12	1.00 − 2	4.36 − 2	7.03 − 3	5.60 − 2	—	3.28 − 2	—
13	4.65 − 3	4.40 − 2	7.03 − 3	1.66 − 2	—	3.28 − 2	—
14	2.15 − 3	4.36 − 2	7.03 − 3	9.13 − 2	—	3.28 − 2	—
15	1.00 − 3	4.36 − 2	7.03 − 3	8.32 − 2	—	3.28 − 2	—
16	4.65 − 4	4.40 − 2	2.24 − 3	1.10 − 1	—	3.16 − 2	—
17	2.15 − 4	4.36 − 2	2.24 − 3	4.93 − 2	—	2.80 − 2	—
18	1.00 − 4	4.36 − 2	2.14 − 3	4.40 − 2	—	2.62 − 2	—
19	4.65 − 5	4.40 − 2	2.14 − 3	3.48 − 2	—	2.38 − 2	—
20	2.15 − 5	4.36 − 2	2.14 − 3	2.55 − 2	—	2.14 − 2	—
21	1.00 − 5	4.36 − 2	4.58 − 4	2.40 − 2	—	1.90 − 2	—
22	4.65 − 6	4.40 − 2	4.58 − 4	1.45 − 2	—	1.55 − 2	—
23	2.15 − 6	4.36 − 2	4.58 − 4	9.07 − 3	—	1.31 − 2	—
24	1.00 − 6	4.36 − 2	4.58 − 4	4.99 − 3	—	9.53 − 3	—
25	4.65 − 7	4.40 − 2	4.58 − 4	1.50 − 3	—	2.38 − 3	—
26	2.15 − 7	5.67 − 2	8.54 − 5	—	—	1.18 − 3	—
Sum	—	1	1	1	1	1	1

[a] The number of neutrons in the given energy group is listed.
[b] Read $a − b$ as $a \times 10^{-b}$.

Characteristics of the shield

Geometry of the shield

Since programs for calculating the characteristics of a neutron field in a one-dimensional geometry are used widely, and since there is a well-developed computational apparatus for analyzing the sensitivity of calculated results to the initial parameters of the problem for a one-dimensional geometry, all the benchmark calculation models listed in Table V were studied in a one-dimensional geometry. The homogeneous or heterogeneous shield layouts consisted of plane (P) or spherical (S) barriers. In the case of a plane shield, the radiation from an infinite plane source was incident on the inner surface of the shield. The detector was on the outer side of the barrier, directly on the surface. In

Table VII. Thicknesses in centimeters of heterogeneous shields of stainless steel, sodium, and iron in benchmark problem 8.

		Shield layout									
Layer	Material	1	2	3	4	5	6	7	8	9	10
1	Stainless steel	47	47	47	47	47	47	47	47	47	47
2	Sodium	155	309	309	309	309	309	460	460	460	460
3	Iron	0	0	15	31	46	62	0	15	31	41

		Shield layout									
Layer	Material	11	12	13	14	15	16	17	18	19	20	21
1	Stainless steel	47	47	31	31	31	31	31	31	31	31	31
2	Sodium	460	460	460	460	460	460	460	460	460	460	460
3	Iron	51	62	0	15	31	41	51	62	72	82	90

the case of a spherical shield, a cavity at the center of the shield held the isotropic point source; the detector was placed either on the outer surface of the sphere[14,109,113,114] or at some distance from it.[113,114] The benchmark problems modeling the actual shield layout for reactors were formulated in an infinite one-dimensional cylindrical geometry (C) for the pressurized water reactor and in the form of a multilayered spherical shield for the fast breeder and fast-neutron reactors.

Shielding material

In most of the studies the shielding materials have been iron and sodium, since these materials are widely encountered in designs for fast-neutron reactors. The thicknesses of these materials have been chosen close to those in designs: from 0.2 to 1 m for iron and from 1.5 to 5 m for sodium. In problems Nos. 1–7, the densities of the iron and the sodium were assumed to be 7.87 and 0.97 g/cm^3, respectively.

The heterogeneous layouts consisted of alternating layers of sodium and iron; as in the shields being designed, the iron was the last layer. In benchmark problem 6, the shields consisted of two layers: a 4.6 m thickness of sodium, simulating a tank with a heat exchanger for a fast-neutron reactor, and 0.5-m and 0.9-m thicknesses of iron.

In benchmark problem 8, a 21-layer shield was studied. This shield consisted of a combination of layers of stainless steel (the closest to the source), sodium, and carbon iron. The thicknesses of the layers in these shields are listed in Table VII.

In benchmark problem 9, a study was made of a multilayer shield consisting of alternating barriers of steel, 0.16 m thick, and graphite, 0.2 m thick, with a total thickness of 0.88 m.

The benchmark calculations on the shields of typical thermal pressur-

Table VIII. Physical zones of the benchmark calculation models of real shields.

Calculation model 10, PWR (cylindrical geometry)			Calculation model 11, FBR (spherical geometry)			Calculation model 12, fast-neutron reactor (spherical geometry)		
Zone	Order	Outer radius of zone, cm	Zone	Order	Outer radius of zone, cm	Zone	Order	Outer radius of zone, cm
Core	1	172.5	Source	1	236.5[a]	Core	1	60
Reflector	2	175.0	Lateral shield	2	416.5		2	62.5
First water layer	3	210.5	Sodium tank	3	916.5	Screen	3	102.5
Jacket	4	218.5				Shield	4	147.5
Second water layer	5	250.0	Heat exchanger	4	966.5		5	222.5
Austenite cladding	6	250.6	Sodium of secondary loop	5	1016.5		6	312.5
							7	327.5
Reactor wall	7	275.6					8	357.5
Concrete shield	8	475.6						

[a] Source thickness is 0.01.

ized water reactors and fast-neutron breeders have been used to develop actual designs for the lateral shielding of a pressurized-water power reactor and a fast-neutron reactor (SNR).

In the Soviet Union, a spherical model of a lateral shield for a fast-neutron reactor has been proposed as this benchmark problem.[29]

Tables VIII and IX show the dimensions and compositions of the zones of these computational models.

Characteristics of the functionals which are calculated

A differential characteristic of the radiation field which can be found through a solution of the transport equation, and whose sensitivity to the parameters of the problem has been studied in several places, is the spatial-energy neutron flux density $\varphi(x,E)$.

In the design of a shield, however, it is frequently sufficient to know certain functionals of this quantity: the heat evolution in the shield (HE), the

Table IX. Zone compositions in the benchmark calculation models for real shields (10^{21} cm^{-3}).

Element or nuclide	Calculation zone of PWR			Calculation zone of FBR					Calculation zone of fast-neutron reactor							
	Core (1)	Reflector, jacket, cladding (2,4,6)	Wall (7)	1st and 2nd water layers (3,5)	Concrete shield (8)	Source and lateral shield (1,2)	Tank and 2nd loop (3,5)	Heat exchanger (4)	1	2	3	4	5	6	7	8
238U	6.3037	—	—	—	—	—	—	—	6.3	5.9	13.5	—	—	—	—	—
235U	2.1040 − 1[a]	—	—	—	—	—	—	—	1.7	2.3	—	—	—	—	—	—
Zr	4.3246	—	—	—	—	—	—	—	—	—	—	—	—	—	—	—
Ni	—	8.5357	—	—	—	4.23	—	1.20	2.4	2.4	2.1	4.0	6.2	1.4	1.4	1.4
Fe	9.8235 − 1	6.4478 + 1	8.465 + 1	—	—	32.00	—	9.06	13.2	13.2	11.4	29.6	45.1	10.3	10.3	10.3
Mn	—	1.1187	—	—	—	—	—	—	—	—	—	—	—	—	—	—
Cr	—	1.6913 + 1	—	—	—	8.60	—	2.43	3.3	3.3	2.9	8.6	13.1	3.0	3.0	3.0
Ca	—	—	—	—	6.6115	—	—	—	—	—	—	—	—	—	—	—
Si	—	—	—	—	9.4350	—	—	—	—	—	—	—	—	—	—	—
Al	—	—	—	—	2.4553	—	—	—	—	—	—	—	—	—	—	—
Na	2.7154 + 1	—	—	2.5278 + 1	4.7751 + 1	10.45	22.23	18.90	7.8	7.8	4.2	11.0	5.3	4.7	4.7	18.3
O	—	—	—	—	—	—	—	—	16.1	16.5	27.1	—	—	48	46	—
C	—	—	—	—	—	—	—	—	—	—	—	—	—	—	—	—
B	—	—	—	—	—	—	—	—	—	—	—	—	—	—	1.0	—
H	2.8226 + 1	—	—	5.0556 + 1	4.4126	—	—	—	—	—	—	—	—	—	—	—

[a] Read a − b as a×10^{-b}.

absorbed dose rate (P), the equivalent dose rate (P_{equ}), the activation of the shield (AS), the radiation damage in the shield materials (RD), the neutron flux density φ, etc. These are the functionals which were used, in various combinations, for the analysis in these benchmark problems. The functionals which were used in Refs. 109, 110, and 115 were the flux density (φ_F) of fast neutrons with energies $E \geqslant 1.4$ MeV (a measure of the radiation damage, in a first approximation), the equivalent dose rate P_{equ}, which is a measure of the biological effect of the radiation, and the response ($\varphi_{1/v}$) of a detector with a $1/v$ neutron detection efficiency, where v is the neutron velocity. Such a detector may be thought of, in a first approximation, as a device detecting the shielding activation rate. In benchmark problem 8 the detector was a Bonner detector, consisting of a spherical counter 5.08 cm in diameter filled with gaseous BF_3 and surrounded with a cadmium-covered polyethylene sphere with a wall thickness of 2.54 cm. This detector detects neutrons with energies from 1 eV to 100 keV with an essentially uniform efficiency.

All the detectors were assumed to be isotropic point detectors in the calculations.

In benchmark problems 9–11, modeling real reactor shields, the following calculation functionals were selected: for the pressurized water reactor, (1) the radiation damage (RD) from neutron radiation in the reactor vessel, (2) the equivalent dose rate P_{equ} at the outer surface of the concrete shield, and (3) the heat evolution (HE) due to γ radiation in the concrete shield; for the fast breeder, (1) the activation of sodium (AS) in the secondary coolant loop and (2) the radiation damage (RD) in the iron in the region of the sodium-iron shield.

Characteristics of the calculation software

Calculation methods and programs

The numerical methods presently used most frequently for solving the kinetic equation are the discrete-ordinates method (DO) and the variational-difference methods (VDM), which are the best-developed methods for solving problems in a one-dimensional geometry. Software implementations of the discrete-ordinates method are the ROZ programs and the American ANISN and DTF-IV. The variational-difference method has been implemented in the ROZ-11 program for the BÉSM-6 computer. The ANISN program is the foremost program in non-Soviet research on the sensitivity of results to the input parameters of the problem in a one-dimensional geometry. It is being used to solve the basic and adjoint neutron-transport equations. The authors of the present book have used the ROZ-11 program (Sec. 2.2) for this purpose. In several of the benchmark calculation problems on sensitivity, especially where the direct-substitution (DS) method has been used, a moment method (MM) (problem 3) and semiempirical diffusion-extraction (DE) methods have been used. These methods have been implemented in the SABINE and REDIFFUSION programs.

Approximations

Various approximations can be used to describe the angular dependence of the solution of the kinetic equation and of the differential scattering cross section, which appears in the transport equation. With increasing order of the approximation, the calculations become more laborious, requiring more computer time. It is thus natural to attempt to evaluate which approximation order is sufficient to obtain an acceptably accurate result.

The order of the approximation of the angular dependence of the solution of the transport equation in the calculations by the ROZ-11 program was taken to be constant, P_9; in calculations by the ANISN program, the approximations of the angular dependence of the solution ranged from S_6 to S_{12} in different problems. Approximations P_0 through P_5 have been used to describe the angular dependence of the differential scattering cross section.

Cross section libraries

In accordance with the basic purpose of sensitivity analysis—evaluating the sensitivity of results to the total and partial neutron-matter interaction cross sections—basic calculations have been carried out with various systems of constants which are widely used in various countries in the practical design of reactor shielding.

Calculations have been carried on in the Soviet Union in a 26-group approximation on the basis of the BNAB-78 library of microscopic cross sections.[89] The data of this library were processed by the ARAMAKO-2F program.[26]

In many of the benchmark problems which have been carried out by European investigators, a 100-group system of neutron cross sections, the EURLIB system (which ignores the resonant self-shielding of the cross sections), has been used along with its fewer-group modifications constructed on the basis of the ENDF/B-III and ENDF/B-IV neutron-data libraries. In several of the problems, a study has been made of the sensitivity of the results to systems of constants differing in the averaging spectrum.

In many studies, the benchmark problems have been solved with the help of the neutron-data libraries ENDF/B-III, ENDF/B-IV, UKNDL, and DLC-2D.

Characteristics of the sensitivity-analysis software

Calculation method and program

One method which is presently being used to solve sensitivity-analysis problems in a one-dimensional geometry is the direct-substitution method, which is used primarily for comparing various systems of constants; for estimating the effect of the source specification function, the order of the angular approximation of the solution, and the scattering cross section, and for estimating the applicability of the results of linear perturbation theory. A second method is based on a generalized perturbation theory (GPT) and is implemented, for example, in the ZAKAT programs (Sec. 2.2) and the SWANLAKE programs for calculating the relative sensitivities of the results to the neutron-matter interaction cross sections.

In the discussion below we will frequently make use of the notation introduced in Table V to summarize the benchmark calculation problem, by specifying all or some of the characteristics of the problem between dashes: the problem number—the characteristics of the source—the characteristics of the shield—the functional being calculated—the characteristics of the calculation software—the characteristics of the sensitivity-analysis software—references.

Chapter 3
Errors in the results of shielding calculations and sensitivity of the results to the input parameters

1. Methodological errors

The mathematical methods for solving the transport equation have now been developed to the extent that they may be regarded as exact in the limit in which we let the order of the approximation of the solution go to infinity (at a sufficiently high-order approximation of the solution). As the order of the approximation increases, the calculations become more laborious, and the rate at which the solution converges on the exact solution may differ substantially in different methods. It is thus common to use approximate methods for estimates. Where necessary, the results of these calculations can be modified by appropriate corrections or correction factors, determined by comparing the results with the results of more accurate calculations.

In this section we examine the methodological errors (a systematic error component) in shielding calculations which stem from the approximations of the methods used to solve the radiation transport equation. The methodological errors of the Monte Carlo method go beyond the scope of this section; we simply note that the error of that method also has a random component, which can be reduced to a comparatively small size by increasing the number of case histories of particles to the necessary level.

The error in the actual method used for the shielding calculation on the basis of a solution of the transport equation stems from the choice of (a) the order of the approximation of the spatial dependence of the solution (the spacing of the spatial mesh), (b) the order of the approximation of the angular dependence of the solution (angular quadrature or the order of the approximation in the spherical-harmonic method), (c) the approximation of the scattering characteristic of the radiations, and (d) the solution algorithm (the numerical scheme, the iterative scheme, etc.).

When the relatively rigorous calculation methods based on a solution of the transport equation in a higher-order approximation are used, the error in the results is determined primarily (at least for one-dimensional calculations) by the error in the radiation-matter interaction cross sections which are used, not by the error of the method itself. This can be seen from Fig. 8, for example,

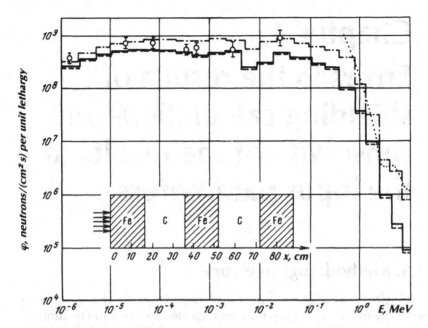

Figure 8. Energy distribution of the neutron flux density in a composite steel-graphite shield at a distance $x = 42.5$ cm. Circles and dotted line, experimental data; solid and dot-dashed lines, calculations by the ROZ-5 program with the DLC-2D and BNAB-64 cross sections, respectively; dashed line, calculations by the ANISN program with the DLC-2D cross sections. The geometry of the problem is shown at the lower left.

which compares the results calculated in Ref. 119 by the ROZ-5 program[129] in the $2D_7P_5$ approximation of the method of characteristics and by the ANISN program[130] in the $S_{12}P_5$ approximation of the S_N method. The neutron energy distributions were calculated in a heterogeneous steel-graphite shield 88 cm thick (calculation model 9, Table V). The discrepancy between the results does not exceed 20% when one system of cross sections, the DLC-2D system, is used. When a switch is made to the BNAB-64 constants, the results of the calculations change by a factor of about 2 at resonance neutron energies and by a factor of up to 8 for fast neutrons (the difference is particularly large at energies above 4 MeV). The calculations with the BNAB-64 constants are in far better agreement with the experimental data. It should be kept in mind, however, that the methodological errors of multidimensional calculations can be quite significant.

Discrete-ordinates method

The methods of this group which are used most commonly in solving the forward and adjoint problems in one and two dimensions are the S_N and DS_N Carlson methods and the method of characteristics.[129]

When these methods are used to solve the boundary-value problems which are characteristic of shielding calculations, a difference mesh is introduced along the angular and spatial variables.

The spatial-angular flux density $\varphi(\mathbf{r},\Omega)$ at the nodes of the difference cell and its average values over the angular and spatial variables in the cell are found by solving a system of algebraic equations which approximate the given boundary-value problem.

In the S_N and DS_N methods, this system of algebraic equations is found by integrating the transport equation along the angular and spatial variables within the difference cell. In the method of characteristics, this integration is carried out along certain special directions: the characteristics.

These methods have been modified to incorporate additional assumptions regarding the dependence of the solution and of the right-hand side (the source function) on the spatial and angular variables within the cell.

We denote by $\varphi_{xp}(\mathbf{r}_k,\Omega_n)$ the solution of the corresponding algebraic system of equations for the finite-difference grid $(\mathbf{r}_k,\Omega_n)$, where $k = 1,...,K$; $n = 1,...,N$. The error of the finite-difference approximation of the problem is then determined by the difference $\varphi_{xp}(\mathbf{r}_k,\Omega_n)$ and $\bar\varphi(\mathbf{r}_k,\Omega_n)$, where $\bar\varphi(\mathbf{r}_k,\Omega_n)$ are the average values of the exact solution $\varphi(\mathbf{r},\Omega)$ in the cell or the values at the nodes of the mesh according to the finite-difference solution found. In a numerical study of the error, the $\bar\varphi(\mathbf{r}_k,\Omega_n)$ are usually chosen to be the values of the finite difference solution $\varphi_{xp}(\mathbf{r}_k,\Omega_n)$ found for as fine a mesh as possible.

For simplicity we consider the case of a plane geometry, for which we can write the transport equation in the form

$$\mu\, d\varphi/dx + \Sigma(x)\varphi(x) = S(x). \tag{3.1}$$

An integration of this equation over the cell of the difference mesh, $\Delta x_k = x_{k+1/2} - x_{k-1/2}$, yields

$$\mu(\varphi_{k+1/2} - \varphi_{k-1/2}) + \Sigma_k\, \bar\varphi_k \Delta x_k = \bar S_k \Delta x_k. \tag{3.2}$$

Here

$$\left.\begin{aligned}
\psi_{k+1/2} &= \varphi(x_{k\pm1/2}); \\
\bar\varphi_k &= \frac{1}{\Delta x_k}\int_{x_{k-1/2}}^{x_{k+1/2}} \varphi(x)dx; \\
\Sigma_k &= \Sigma(x_k); \quad x_k = (x_{k-1/2} + x_{k+1/2})/2.
\end{aligned}\right\} \tag{3.3}$$

When the S_N method is used, it is necessary to use some additional equations which relate the flux density at the center of a cell in phase space, φ_k (or the average over the cell $\bar\varphi_k$), to the flux density at the cell boundaries, $\varphi_{k-1/2}$ and $\varphi_{k+1/2}$.

One of the most widely used schemes is the diamond difference scheme,[131] which employs the equation

$$\bar\varphi_k = (\varphi_{k-1/2} + \varphi_{k+1/2})/2. \tag{3.4}$$

From Eqs. (3.3) and (3.4) we find

$$\varphi_{k+1/2} = \frac{2-\delta_k}{2+\delta_k}\varphi_{k-1/2} + \frac{2\delta_k}{2+\delta_k}\frac{\bar S_k}{\Sigma_k}, \tag{3.5}$$

where $\delta_k = \Sigma_k \Delta x_k/\mu$ is the optical distance along the trajectory in the kth cell. We thus see that even if $\varphi_{k-1/2}$ and $\bar S_k$ are positive, this equation can lead to negative values of the angular flux density and sometimes of the total

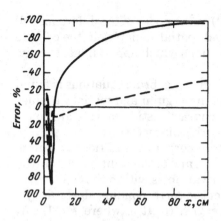

Figure 9. Error in the flux density of γ radiation with an energy of 9 MeV in iron shielding (Ref. 132) for spatial mesh spacings $\Delta x = 2$ cm (dashed line) and $\Delta x = 5.16$ cm (solid line), found as the difference from the results for $\Delta x = 0.25$ cm.

flux at $\delta_k > 2$. This fact is one of the basic reasons for the error of the S_N method, which leads in certain cases to large oscillations of the spatial distribution of the flux density.

The error of calculations using the diamond-difference scheme turns out (for small spacings Δx) to be proportional to the second power of the step of the spatial mesh, Δx; i.e., in each step we have $\Delta \varphi \sim - (\Delta x)^3$, so that for a layer of shielding we have $\Delta \varphi = K \Delta \varphi \sim - (\Delta x)^{-1}(\Delta x)^3 = - (\Delta x)^2$. To illustrate the extent of the error of the calculations for various steps Δx we show in Fig. 9 the dependence of this error on the distance from the source in calculations in the $S_8 P_3$ approximation for the transport of γ rays with a reactor spectrum through a layer of iron 100 cm thick by the DTF-IV program.[132] We see that when large spacings are used the error of the calculation is significant, and at $\Delta x = 5.16$ cm the results on the γ flux density beyond the iron layer are 98.5% lower than the exact solution (which we can take to be the data for $\Delta x = 0.25$ cm), i.e., by a factor of about 60.

As the spacing is reduced, the error decreases: $\Delta \varphi / \varphi \approx 0.3$, 0.07, and 0.01 at $\Delta x = 2$, 1, and 0.5 cm, respectively.

Unfortunately, the conclusion which follows—that it is necessary to reduce the spacing (Δx) of the finite-difference mesh in the calculations in order to improve the accuracy of the calculations—requires an increase in computer time, and this increase is sometimes substantial. In other words, the calculations become unacceptably expensive, and in several cases impossible, because of limited computer speed and memory.

In practice, therefore, many other finite-difference approximations are used. One which is regarded as among the most promising is the exponential method,[133] which makes use of the assumption of an exponential distribution of the flux density within a cell. In this case the following relation can be used

$$\bar{\varphi}_k = (\varphi_{k-1/2} \varphi_{k+1/2})^{1/2} . \tag{3.6}$$

The error of the calculations (a local error) in the exponential approximation also turns out to be proportional to $(\Delta x)^3$ (for small spacings), as in the use of the diamond difference scheme, but the sign is different: $\Delta \varphi_{\bullet\bullet} \sim (\Delta x)^3$. In other words, this approximation overestimates the results. The proportionality factor turns out to be half that for the diamond case. The exponential approxima-

Figure 10. The ratio (κ) of the results calculated for the fast-neutron flux density in water by the ANISN program for various spacings Δx to the results for Δx = 0.125 cm. (a) diamond difference scheme; (b) exponential approximation.

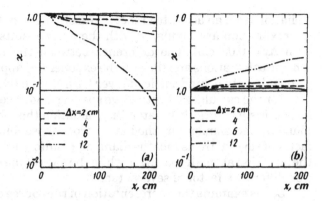

Table X. Relative error (as a percentage) of the calculations of the energy flux density of γ radiation in lead by the S_N method in the ANISN program (Ref. 133).

x, cm	Linear approximation (diamond difference scheme)			Exponential approximation		
	Δx = 0.25 cm	Δx = 1 cm	Δx = 6 cm	Δx = 0.25 cm	Δx = 1 cm	Δx = 6 cm
0.5	− 1.4	+ 7.4	+ 7.4	1.4	+ 5.5	+ 5.5
6.5	− 0.5	− 7.5	− 12.0	+ 0.4	+ 6.1	+ 34.7
11.5	− 0.7	− 10.3	+ 74.1	+ 0.5	+ 7.2	+ 105.2
17.5	− 0.9	− 14.0	+ 637.8	+ 0.6	+ 9.3	+ 250.3
23.5	− 1.2	− 18.3	+ 2 970	+ 0.8	+ 11.8	+ 384.7
29.5	− 1.4	− 22.4	+ 12 340	+ 1.0	+ 14.5	+ 898.8
35.5	− 1.7	− 26.4	+ 50 260	+ 1.2	+ 17.4	+ 1571.4
40.5	− 2.0	− 29.6	+ 184 360	+ 1.3	+ 19.9	+ 2395.4
44.5	− 2.2	− 32.0	+ 295 330	+ 1.4	+ 22.0	+ 2812
49.25	− 2.5	− 36.4	+ 508 510	+ 1.3	+ 23.9	+ 3230

tion with positive sources always leads to positive values of the flux density.

Figure 10 shows results calculated from these two schemes for the flux density of fast neutrons (with $E > 0.82$ MeV) in a water shield 200 cm thick which is being bombarded by an isotropic plane source of fission neutrons. Shown for comparison here are the results of the exact solution (i.e., with a spacing $\Delta x = 0.125$ cm).[134] These calculations were carried out by the ANISN program in the $S_8 P_3$ approximation for a 26-group system of cross sections found from the 100-group DLC-2D library.

It can be seen from Fig. 10(b) that the exponential approximation always overestimates the result; the error of the calculations (for small spacings) is roughly half that of the diamond-scheme calculations, which is rather large. For example, at $\Delta x = 12$ cm the diamond-scheme calculations underestimate the result by 95%, i.e., by a factor of about 20.

The advantages of the exponential approximation increase with increasing size of the spacing (at least for homogeneous shields). The use of the exponential approximation also causes a slight increase in the convergence rate of the iterative process (but the dependence of this rate on Σ_s / Σ_t is stronger).

Table X compares the results of corresponding calculations of the energy flux density of γ radiation in a lead layer 50 cm thick bombarded by the γ radiation from an activated reactor vessel. The results calculated in the S_8P_4 approximation are compared with the exact results (i.e., those for the case with $\Delta x = 0.03$ cm). The difference between the results of the diamond-scheme calculations and those in the exponential approximation for the energy flux density behind the lead layer with $\Delta x = 0.03$ is only 0.05%.

For the calculations with large spacings, however, the error of the calculations is substantially greater for γ radiation than for neutrons. The results found by the diamond method differ from those found with $\Delta x = 6$ cm by a factor of several thousand (the diamond results are too high, in contrast with the situation at small spacings), while the data of the exponential approximation differ by a factor of several tens.

Let us examine the representation of the source distribution function. In the solution of the transport equation by an iterative method, one uses some assumption or other regarding the nature of the distribution of the source $S(x,\mu)$ between nodes $x_{k-1/2}$ and $x_{k+1/2}$. The usual assumption is a linear interpolation:

$$S_{xp}(x,\mu) = a_0 + a_1(x - x_{k-1/2}) \quad \text{for } x \in (x_{k-1/2}, x_{k+1/2}); \tag{3.7}$$

$$\overline{S}_{xp} = (1/\Delta x) \int_{x_{k-1/2}}^{x_{k+1/2}} [a_0 + a_1(x - x_{k-1/2})]dx = a_0 + a_1\Delta x/2. \tag{3.8}$$

The constants a_0 and a_1 can be found by the various schemes through the use of the following assumptions:

(1) Conservation (upon the switch to the finite-difference representation of the source) of the "node" values of the source, S_{k-1} and S_k,

$$a_0 = S_{k-1/2}; \quad a_1 = (S_{k+1/2} - S_{k-1/2})/(x_{k+1/2} - x_{k-1/2}). \tag{3.9}$$

(2) Conservation of $\overline{S}$ and of the gradient of the source. In this case the constants a_0 and a_1 are chosen in such a manner that we have

$$\int_{x_{k-1/2}}^{x_{k+1/2}} S_{xp}(x)dx = \int_{x_{k-1/2}}^{x_{k+1/2}} S(x)dx = \overline{S}\Delta x. \tag{3.10}$$

$$dS_{xp}(x)/dx = (S_{k+1/2} - S_{k-1/2})/\Delta x. \tag{3.11}$$

We then find

$$a_0 = \overline{S} - (S_{k+1/2} - S_{k-1/2})/2; \quad a_1 = (S_{k+1/2} - S_{k-1/2})/\Delta x_k. \tag{3.12}$$

(3) Conservation of $\overline{S}$ and of the left-hand node source, $S_{k-1/2}$,

$$a_0 = S_{k-1/2}; \quad a_1 = 2(S_k - S_{k-1/2})/\Delta x_k. \tag{3.13}$$

(4) Conservation of $\overline{S}$ and of the right-hand node source, $S_{k+1/2}$,

$$a_0 = 2\overline{S}_k - S_{k+1/2}; \quad a_1 = 2(S_{k-1/2} - S_k)/\Delta x_k. \tag{3.14}$$

A quadratic interpolation of the source distribution function is of course more accurate:

$$S_{xp}(x,\mu) = a_0 + a_1(x - x_{k-1/2}) + a_2(x - x_{k-1/2})^2$$
$$\text{for } x \in (x_{k-1/2}, x_{k+1/2}). \tag{3.15}$$

The corresponding average value of the source is

$$\overline{S}_k = a_0 + a_1\Delta x/2 + a_2\Delta x^2/3. \tag{3.16}$$

Figure 11. Error in the calculation of the neutron flux density as a function of the size of the spacing of the spatial mesh (expressed in units of the mean free path). (a) $S(x) = \exp(-x)$; (b) $S(x) = x\exp(-x)$ (Ref. 134). The curves are labeled with the index of the scheme used for the spatial approximation of the source function, as explained in the text proper; 5, quadratic interpolation.

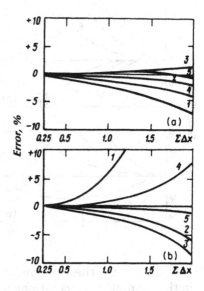

In this case the constants are chosen from the conditions

$$\int_{x_{k-1/2}}^{x_{k+1/2}} dx\,[S_{xp}(x) - S(x)] = 0; \tag{3.17}$$

$$S_{xp,k+1/2} = S_{k+1/2}\,. \tag{3.18}$$

As a result we find

$$a_0 = S_{k+1/2}; \tag{3.19}$$

$$a_1 = 2(3\overline{S}_k - 2S_{k-1/2} - S_{k+1/2})/\Delta x_k\,; \tag{3.20}$$

$$a_2 = 3(S_{k+1/2} - S_{k-1/2} - 2S_k)/\Delta x_k^2\,. \tag{3.21}$$

For a comparison of these interpolation schemes, single-velocity calculations have been carried out for two trial source functions,[134] $x\exp(-x)$ and $\exp(-x)$. This choice of trial functions was based on the well-known fact that in a plane layer the scattered radiation accumulates not far from the source, as described by $x\exp(-x)$, while at a certain distance from the source we find $\exp(-x)$. For these source functions, the flux densities can be found analytically. Figure 11 shows values of the error in the calculations of the flux density for these schemes with $\Sigma_s/\Sigma_t = 0.5$. It can be seen from this figure that the least accurate is the first scheme, which uses the node values of the source. Those schemes (the fourth and the second) which use the values of either $S_{k+1/2}$ and $\overline{S}_k$ or $\Delta S/\Delta x$ are roughly equally accurate. These schemes make it possible to use spacings of up to one mean free path with an error of less than 0.5%. The most accurate of the schemes considered here is the quadratic interpolation, which leads to an error of 0.5% only for a spacing of two mean free paths.

The accuracy class of these schemes has been found analytically for small spacings[135,136] (for a slab shield): second class, i.e., $\Delta\varphi/\varphi \sim (\Delta x)^2$, for the first scheme; third class, i.e., $\Delta\varphi/\varphi \sim (\Delta x)^3$, for the third and fourth schemes; and fourth class, i.e., $\Delta\varphi/\varphi \sim (\Delta x)^4$, for the second and fifth schemes.

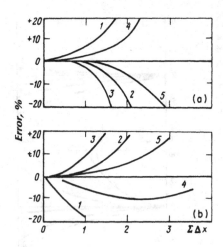

Figure 12. Error in the accumulation factors for γ radiation with an energy $E_0 = 1$ MeV beyond a lead plate 10 mean free paths thick in (a) the reflection problem and in (b) the transmission problem. See Fig. 11 for the rest of the notation.

The error of these schemes for multigroup calculations has been studied in the example of calculations in the S_8 approximation of the transport of γ rays with an energy of 1 MeV in an iron layer with a thickness equal to ten mean free paths.[134] Figure 12 shows data on the error of these calculations in comparison with the calculations with $\Sigma \Delta x = 0.1$. As in the case of the analysis with the trial functions, we see that among the schemes with a linear interpolation the most accurate are schemes 2 and 4. These two schemes make it possible to use a spacing with a width of one mean free path with an error of 1% in the accumulation factor for the transmitted radiation (beyond the shield), in comparison with $\Sigma \Delta x = 0.4$ for a scheme which uses "node" values of the source. As in the single-velocity problem, the quadratic scheme provides the greatest accuracy by virtue of the use of a larger amount of information (two node values and the average value of the source).

Other schemes are also used in practical calculations. In the ONETRAN program,[137] for example, yet another version of linear interpolation (3.7) is used: one which conserves $\bar{S}$, $S_{k+1/2}$, and

$$\int_{\Delta x_k} (x - x_{k-1/2}) S(x) dx.$$

Calculations in the $S_8 P_3$ approximation show that for the passage of neutrons from a 14-MeV source in a concrete shield 300 cm thick the neutron flux density found beyond the shield with a spacing $\Delta x = 5$ cm differs from the corresponding flux density with a spacing $\Delta x = 0.8$ cm by 1.8%.

In this program, discontinuous finite elements are used for a spatial approximation of the flux density.[138] This effect, combined with the use of discontinuous angular finite elements, considerably refines the results, including the results obtained from calculations on shielding with absorbers, voids, and a high-order scattering anisotropy, without the cost of an increased consumption of computer time. It is expected that this approach may also be effective for eliminating ray effects in a two-dimensional geometry (more on this below).

Another way to write the source function, which tends to reduce the

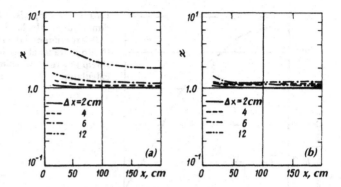

Figure 13. The ratio (κ) of the results calculated for the fast-neutron flux density in water by the PALLAS program for various spacings Δx to the results for $\Delta x = 0.125$ cm. (a) Linear representation of the source function; (b) combination of linear and exponential representations of the source function.

calculation error, is to approximate this function in each cell by a sum of spatial moments of the angular flux[132]:

$$M_k^l(\mu) = \int_{x_{k-1/2}}^{x_{k+1/2}} dx (x - x_{k-1/2})^l \varphi(x, \mu) . \tag{3.22}$$

When this approximation by a polynomial is combined with the use of the weight function $g(x) = \exp(-\Sigma_{\min} x)$, where $\Sigma_{\min}$ is the smallest total cross section in all the groups considered, it becomes possible to substantially improve the accuracy of the calculations when only two spatial moments are used. If four moments are used, the spacing of the spatial mesh can be increased by a factor of about 100. In calculations on the transport of γ radiation with a reactor spectrum in a 1-m thickness of iron, e.g., the switch from the spacing $\Delta x = 0.25$ cm to $\Delta x = 25$ cm increased the calculation error by only about 1%.

Probably the most economical scheme for approximating the source function, and one which significantly improves the accuracy of calculations, including calculations in a curvilinear or two-dimensional geometry, is the scheme used in the PALLAS program,[139,140] which is based on the method of characteristics (a direct integration along the trajectory of the particles). A combination of linear and exponential representations is used for the source function $S(r)$. The exponential representation is used if the drop in the source strength over the cell of the finite-difference spatial mesh is more than 2 or less than 0.5. Otherwise, a linear representation is used. The exponential approximation is more suitable in cases in which unscattered radiation plays an important role; the linear approximation is more suitable when the scattered radiation is dominant.

According to the results of these calculations, if the drop in $S(r)$ in a cell is by a factor ~ 2, the error of the linear representation of the flux density at the center of the cell is $\sim 6\%$, while for a drop by a factor of 3 the corresponding error is $\sim 15\%$: clearly unacceptable.

Figure 13 shows data on the error in calculations of the fast-neutron

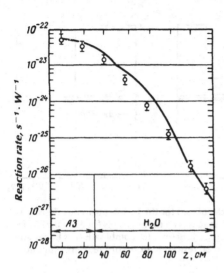

Figure 14. Comparison of the results of two-dimensional calculations by the PALLAS program with experimental data on the activity of a $^{58}Ni(n,p)$ tracer along the height of the lateral shield of a reactor. Points, experimental; line, calculated.

distribution in water[140] obtained by the PALLAS program in the S_{20} approximation with a spacing Δx varied from 2 to 12 cm.

We see that the use of the combined representation of the source function makes it possible to use rather large spacings ($\Delta x = 12$ cm) without significantly increasing the calculation error.

Similar calculations by the PALLAS program have been carried out on the distribution of the γ energy flux density in a lead slab 50 cm thick for 34 γ groups and for an isotropic fission γ source.[140] Analysis of these results shows that the error of these calculations is significantly smaller than the error of calculations using the ANISN program. At $\Delta x = 4$ cm the maximum error is 110% for the case of a linear representation of the source function and 4% for the combination of linear and exponential representations; at $\Delta x = 2$ cm the maximum error is 20% for the linear representation.

Heterogeneous shields may make the situation more complicated. Furthermore, when we go to a curvilinear or two-dimensional geometry we find it necessary to interpolate the solution within each cell, and the error of the method may increase, at least in calculations on heterogeneous shielding.

In an effort to determine the error of the method used in the PALLAS program for two-dimensional (r,z) shielding calculations, a two-dimensional version of this program was used to carry out calculations on an experiment carried out in the JRR-4 water-cooled, water-moderated reactor. Figure 14 compares the calculated results with experimental data on the activity distribution of a $^{58}Ni(n,p)$ tracer along the height of the lateral shield of the reactor. The diameter of the reactor is 43.1 cm. The lateral shield includes graphite (23.2 cm), aluminum (1.5 cm), water (39.2 cm), and iron (20 cm). Measurements were carried out along the vertical in a layer of water beside the inner surface of the iron layer ($r = 85.5$ cm). The calculations were carried out with average spacings Δr, Δz less than 2.5 cm. We see that the calculated results exceed the experimental results by a factor of about two at $z \approx 80$ cm and undergo wavelike oscillations near $z \approx 50$ cm because of ray effects.

Table XI shows the results of a study of the dependence of the calculation

Table XI. Error (%) of two-dimensional calculations on the activity of a ^{58}Ni(n,p) tracer by the PALLAS program.

z, cm	$\Delta r = \Delta z = 5$ cm	$\Delta r = \Delta z = 10$ cm	z, cm	$\Delta r = \Delta z = 5$ cm	$\Delta r = \Delta z = 10$ cm
0	15.4	32.8	100	41.8	72.4
20	17.1	56.2	120	80.0	124.3
40	22.5	55.2	140	54.9	141.3
60	26.6	90.3	160	42.5	125.6
80	39.1	80.6	180	32.1	93.1
			200	14.2	47.4

error on the spacing in the spatial mesh. Shown here are the deviations of the results calculated on the activity of ^{58}Ni(n,p) tracers with various spacings $\Delta r, \Delta z$ at points along the line where the experiment was carried out from the case $\Delta r = \Delta z = 2.5$ cm (z is the distance from the central plane of the core of the reactor).

It can be seen from Table XI that the two-dimensional calculations with large spacings overestimate the results. The spatial distribution of the error is, however, not a monotonic function, as it usually is in one-dimensional calculations. The largest error is seen near $z = 120$ cm; beyond this point, the errors decrease.

In general, this numerical simulation showed that the use of the scheme proposed in the PALLAS program makes it possible to achieve results of satisfactory accuracy even with a significant increase in the mesh spacing (to reduce the amount of computer memory and computer time required), up to $\Delta r = 10$ cm.

Similar results on the error of calculations for the combined representation of the flux density function (a linear representation at distances less than a mean free path from the boundaries of the slab and an exponential representation in the slab) were obtained by the ROZ-5 program,[129] which uses a method of characteristics in plane geometry. In the case of isotropic scattering with $\Sigma_s / \Sigma = 0.9$, the error of the calculation of the radiation flux density transported across a slab 10 mean free paths thick is only 10%–20% at a spacing $\Sigma \Delta x = 3$–4. With a linear anisotropy of the scattering and a spacing $\Sigma \Delta x = 1$, this error decreases to $\sim 0.1\%$. Corresponding calculations with a linear representation of the source function yield results differing by a factor of several units from the exact solution, found with a spacing $\Sigma \Delta x = 0.05$, while the discrepancy increases to a factor of several thousand for large spacings.

Important properties of difference schemes are the positivity (for positive sources S and boundary conditions) and monotonic nature (the absence of oscillations) of the results of the solution of the problems.

For heterogeneous shields, the spatial and angular derivatives of the solution may be discontinuous or have a singularity at the boundaries between zones and at surfaces tangent to them.[141] In the case of nonmonotonic schemes, the result will be large errors on coarse meshes; because of the balanced nature of the DS_N method, these large errors will propagate as oscillations.[142]

Table XII. Error (%) in the fast-neutron flux density calculated in the $2D_N P_3$ approximation by the ROZ-2 program (angular Gaussian quadrature).

E, MeV	x, cm	$N = 1$	$N = 2$	$N = 3$	$N = 5$	$N = 7$
$\geqslant 2.5$	10	32.5	2.4	-0.31	-0.12	-0.05
	40	-55.0	-9.0	-0.62	0.19	0.15
	50	-60.6	-19.1	0.21	0.20	0.19
	80	-84.8	-35.5	-2.13	-0.14	-0.21
$\geqslant 1.4$	10	21.8	1.1	-0.16	-0.03	0.02
	40	-57.3	-97.7	0.55	0.15	0.13
	50	-63.1	-17.5	0.37	0.15	0.14
	80	-85.1	-35.0	-2.08	-0.08	-0.13
0.1–1.4	10	15.6	0.33	-0.07	-0.01	0.01
	40	-60.0	-9.58	0.44	0.05	0.04
	50	-60.3	-5.56	0.29	0.05	0.03
	80	-98.6	-31.9	-1.68	-0.14	-0.15

In the versions of the DS_N method described above, the oscillation problem was not finally resolved. Certain improvements in the solution can be made by making special corrections for negative values of the fluxes (e.g., setting them equal to zero). In the new MDS_N method,[142,143] a special monotonic scheme is used for calculations for cells with singularities in the solution. Let us consider angular quadratures. To demonstrate the extent to which the choice of the angular mesh affects the errors in shielding calculations, we show in Table XII data on the error in the fast-neutron flux density in the iron-water shield of a water-cooled, water-moderated reactor. The calculations[144] were carried out by the ROZ-2 program by a method of characteristics, in the $2D_N P_3$ approximation,[129] for a shield consisting of 10 cm of iron, 30 cm of water, 10 cm of iron, and 30 cm of water directly beside the core.

We see that, for an attenuation of the neutron flux density up to 10^5, only at $N \geqslant 5$ is it possible to keep the error in the neutron flux density below 20%. At a lower attenuation, the $2D_3$ approximation can be used.

Numerous calculations by the ROZ-5 program have shown that the $2D_5$ approximation is usually the best for calculating the angular and total flux densities with satisfactory accuracy without a substantial expenditure of computer time.

Figure 15 illustrates the behavior of the error in calculations in a two-dimensional (r,z) geometry as a function of the angular quadrature. Shown here are the results of single-velocity calculations carried out in the DS_N approximation with $N = 2, 4$, and 8 (Ref. 145) of the spatial distribution of the neutron flux density in a semi-infinite, homogeneous medium with isotropic scattering (with $\Sigma_s / \Sigma_t = 0.4$) for a unidirectional point source. Also shown here are the results of Monte Carlo calculations (with 60 000 case histories). We see that the calculation error increases with distance from the beam axis, particular for the DS_2 approximation at small values of z. The error of the DS_4 calculations reaches 100%, and that of the DS_2 calculations becomes even larger. It can thus be concluded that the error of the two-dimensional DS_N calculations is quite large at $N < 8$.

Figure 15. Results calculated on the spatial distribution of the neutron flux density in a semi-infinite medium for a unidirectional point source. Circles, Monte Carlo method; solid line, DS_8 approximation; dashed line, DS_4 approximation; dotted line, DS_2 approximation. The values of the flux density averaged over the calculation cells (Δz, Δr) are shown for the DS_N calculations.

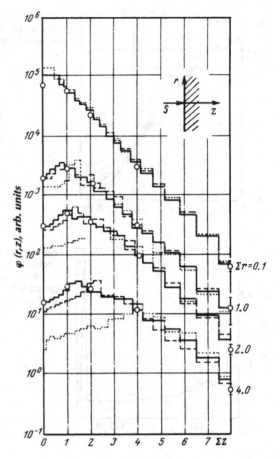

To illustrate the error of two-dimensional calculations of the angular distributions of the flux density, we show in Fig. 16 the results of such calculations for fast neutrons in toroidal (r, θ_0) geometry. Shown here are angular distributions of the neutrons in a cylindrical steel wall with an inside radius of 210 cm, at a distance of 30 cm from this radius.[146] The source is a filamentary source in the form of a ring with a radius of curvature $R = 625$ cm; the energy of the source neutrons is 14 MeV. We see that the DS_6 approximation has a rather large error over the entire range of angles. The maximum error of the DS_8 approximation is 4% (in comparison with the calculations in the DS_{16} approximation).

We would obviously like a criterion for choosing the angular quadrature correctly. Such a criterion[147] can be constructed by requiring that the quadrature selected must give an accurate description of the characteristic for neutron scattering by light nuclei, with the piecewise-linear nature of this indicatrix being taken into account (Fig. 17) and with nonzero values only in certain intervals of the argument (the "carrier"). A correctly chosen angular quadrature should make it possible to describe all the nonvanishing values of the intergroup transitions $\sigma^{i \to j}$ ($\mu_l \to \mu_t$), where μ_l and μ_t are quadrature nodes. We know that the structure of the transitions and thus the correspond-

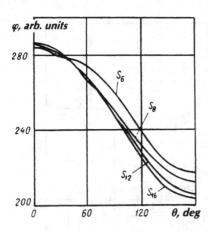

Figure 16. Convergence of DS_N calculations in a two-dimensional toroidal geometry. The angle θ is measured from the direction specified by the angle θ_0.

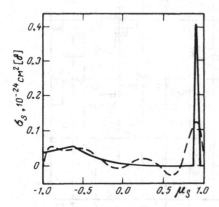

Figure 17. Angular dependence of the cross section for the scattering of neutrons with energies of 3.01–3.33 MeV for water with moderation in the group interval 2.47–2.73 MeV. Solid line, exact; dashed line, P_8 approximation.

ing exact angular quadrature depend on the structure of the energy groups. Table XIII shows data on the percentage of successes of Gaussian quadratures of various orders N for the angular scattering of neutrons by light nuclei[147] for the top 20 groups of the DLC-2D library (the width of the energy groups in the fast-neutron region is ~ 1 MeV; the boundaries of some of these energy intervals are given in Fig. 18, which shows the angular distributions of neutrons transported by and reflected by a layer of water). The "percentage of successes" here means that fraction of all possible intergroup transitions which are described by the given quadrature by no less than three angular transitions (i.e., the fraction of cases in which at least three points fall in the "carrier"). A success percentage of 100% would mean that for the given group structure and the angular quadrature any intergroup transition will be described by at least three angular transitions ($\mu_i \rightarrow \mu_t$). We see that the description of the characteristic is most complex in the limit $A \rightarrow 1$, especially for the low-order quadratures.

It should be noted that there is also a trend opposing that described above, of an increase in the calculation accuracy through a transition to asymmetric angular quadratures, "crowded" in the region $\mu \rightarrow 1$ for a description as accurate as possible of intergroup transitions. Such quadratures hold

Table XIII. Percentage of successes of a Gaussian quadrature for the angular scattering of neutrons by nuclei with mass number _A_ for the DLC-2D library.

N	Mass number			
	1	2	3	4
4	0	0	0	0
6	0	0	0	0
8	0	0	36.5	36.5
10	0	39.0	39.0	89.7
12	0	40.6	91.5	91.5
16	42.6	93.6	93.6	100
32	100	100	100	100

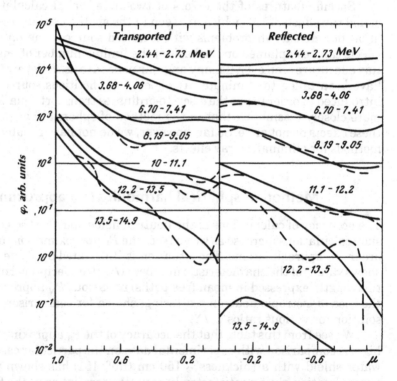

Figure 18. Comparison of the angular distribution of the fast-neutron flux density calculated in the P_8 approximation of the scattering characteristic (dashed line) with the results of exact calculations (solid line) in the problem of the reflection and transport of fission neutrons by a layer of water 10 cm thick.

promise for describing forward-peaked scattering characteristics (neutron scattering by hydrogen or γ scattering).

In addition to the components of the methodological error of the calculations which we have discussed above, which stem from the choice of spatial and angular meshes and the choice of approximation scheme, there are other methodological errors, which are less important; an example is the error of the iterative scheme. Although the latter error can be adjusted by choosing

Table XIV. Values of the characteristic number $\nu(P_N)$ for isotropic scattering ($\Sigma_t = 1$).

Σ_s / Σ_t	P_1	P_2	P_3	Exact value of ν
0.9	0.547 72	0.527 05	0.525 56	0.525 43
0.5	1.224 74	1.035 10	0.988 75	0.957 50
0.3	1.449 14	1.160 24	1.080 12	0.997 41

the criterion for the convergence of the iterative process, it does depend on the number of iterations and on the parameters selected for the convergence of the iteration process, especially when some method or other is used to accelerate this convergence.

Specific features of the errors of two-dimensional calculations are so-called ray effects,[148–150] which are seen in the spatial oscillations of the radiation flux density in problems with localized sources. The nature of these effects can be explained on the basis of the limited number of angular nodes. Ray effects are seen even if many angular nodes are used. Several approaches have been taken to eliminate ray effects; e.g., fictitious sources have been introduced in order to put the discrete-ordinates transport equation in a form more closely resembling that in the method of spherical harmonics, where these effects do not arise. So far, however, we do not have reliable and simple methods for eliminating ray effects.

Method of spherical harmonics (P_N approximation)

The accuracy of calculations of the neutron fields and also the complexity of such calculations increase as we go from the P_1 approximation to the higher-order P_N approximations. The situation is illustrated by Table XIV, which shows values of the characteristic number $\nu(P_N)$ (the reciprocal of the relaxation length, expressed in mean free paths) in various P_N approximations for the case of isotropic neutron scattering; shown for comparison is the exact solution for various ratios Σ_s / Σ_t.

We see from this table that the accuracy of the P_3 approximation is quite high. A comparison with experimental data on the particular case of a metal-water shield with a thickness ~ 150 cm (Ref. 151) has shown that the P_3 approximation gives a satisfactorily accurate description of the fast-neutron fluxes in one-dimensional problems.

The error of the P_N approximations is determined in each particular case by the anisotropy of the angular distribution of the neutron flux, which depends primarily on the ratio Σ_s / Σ_t. The diffusion of neutrons in weakly absorbing materials is described well even by the P_1 approximation, which is the same as the diffusion approximation in one-velocity problems. In hydrogenous materials or in a strongly absorbing medium, the error of the low-order P_N approximations increases substantially. Another factor which influences the accuracy of an approximation is the degree of anisotropy of the neutron scattering, whose relationship with the anisotropy of the angular distribution of the neutron flux is quite obvious.

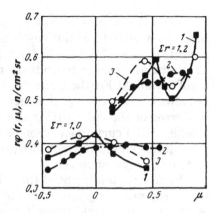

Figure 19. Angular distribution of the neutron flux density in a two-layer spherical region. 2, S_{10} approximation; 1, method of characteristics ($2D_{18}$ approximation); 3, ROZ-11 ($N = 11$, $K = 50$).

Clearly, if the scattering is highly anisotropic, a higher-order approximation will be required to achieve a given accuracy than in the case of a slight anisotropy. Even for fast neutrons in heavy materials, however, where the removal of neutrons by inelastic scattering is substantial, and the elastic scattering is also very anisotropic, the P_3 and P_4 approximations are sufficient in practice for calculations on the distribution of the neutron flux density at distances out to about 150 cm.

The convergence of the P_N approximations, i.e., the decrease in the discrepancy between their solutions and the exact solution, with increasing N is far poorer for the angular neutron flux density $\varphi(x,\mu)$ than for the flux density $\varphi(x)$.

It should be noted that the error of the diffusion-approximation calculations increases when we go from one-dimensional to two-dimensional problems. For example, the error in results of diffusion calculations in an infinite medium with $\Sigma_s/\Sigma_t = 0.9$ at a distance $\Sigma_t x = 9$ in comparison with an exact calculation by the discrete-ordinates method is 7%, while in a bounded cylindrical medium with a radius of 2 mean free paths this error is 45% (Ref. 152).

Variational-difference methods

To illustrate the error of the methods of this group, which synthesize a solution of the problem on the basis of variational principles, we consider the variational-difference method with finite elements which is implemented in the ROZ-11 program (Sec. 2.2).

This method describes quantities integrated over angle quite well. Its accuracy is poorer in the case of angular distributions. For example, the only way to describe the angular distributions in the case of a heterogeneous spherical shield, distinguished by singularities in the angular distributions [discontinuities of the derivatives $\varphi(\mu)$ along lines tangent to the interfaces between layers] is to use the method of characteristics. Figure 19 shows some illustrative angular distributions with singularities for a two-layer shield (the first zone has a radius of 1 mean free path with $\Sigma_s/\Sigma_t = 0.5$; the second has a thickness of 4 mean free paths with $\Sigma_s/\Sigma_t = 0.9$).

We will end this section with some brief conclusions which are important in practical shield calculations. It is generally a quite complicated problem to identify the size of the methodological component of the error of calculations. The calculation error depends strongly on the layout, geometry, and materials of the shield. Furthermore, increasing the order of an approximation does not always make the solution more accurate. In some cases, the methodological error of the calculations can be exceedingly large, especially for real shields with a curvilinear geometry. In this section we have described only some quite general characteristics of the various methods, primarily to illustrate the basic tendencies.

From the standpoint of those who are doing practical calculations and those who are using the results of these calculations, each calculation tool (each program) should make it possible to carry out detailed studies, in various versions, of the convergence of the approximations of the given method, of the iterative process, and of the effect of the choice of the spatial mesh and the angular quadrature. It should also be possible to identify structural features of various types, e.g., ray effects. For many methods there may be some general requirements, e.g., a recommendation to use a finer angular mesh and a coarser spatial mesh for groups of fast neutrons than for moderating and thermal neutrons. However, there may be specific features in the various methods.

These studies should be carried out not by comparing the results with experimental data, since the picture is always blurred by the effect of errors in the nuclear constants in such comparisons, but by means of special numerical simulations with the same system of constants: a comparison with special, high-accuracy reference calculations.

2. Sensitivity of the results to the parameters of the radiation source and of the detector response function

The result of a calculation is affected by the following parameters of the source: (1) its geometry, (2) its strength, (3) the energy distribution of the emitted particles, and (4) the angular distribution of the emitted radiation. Those inaccuracies in the calculated result which stem from inaccuracies in the specification of the source strength can be determined easily if the latter are known, since they are conveyed completely in the result of the calculation.

A shield designer usually specifies the geometry of the source quite accurately. The question of the effect of the source geometry on the result of a calculation may arise if a significant simplification of the geometry is assumed or if a study is being made of the use of characteristics (such as the attenuation of radiation in the shield) obtained in one geometry of the source to solve problems in another geometry.

The method of direct substitution can be used to estimate the effect of the source geometry on the result which will be found from a calculation.

The angular distribution of the source radiation is frequently modeled by a $\cos^m\theta$ dependence.

Since the source distribution function is not a linear function of the parameter m, it is inadvisable to use perturbation theory to analyze the sensitivity of the characteristics of the neutron field to the parameter m of the angular distribution of the source radiation. The method of direct substitution may be an acceptable analysis method in this case.

In an analysis of the effect of the distribution function of the source neutrons on the result of a calculation there is a linear dependence of the source distribution function $S^i(x,\mu)$ on the parameters n_i, which characterize the yield of neutrons in energy groups i. In this case, therefore, the second term in Eq. (1.29) can be used for calculations of the relative sensitivity R to the source specification function. For the plane, unidirectional sources considered in the models for calculations Nos. 1 and 6 (Table V), this expression can be transformed with the help of

$$S^i(x,\mu) = \delta(x)\delta(\mu - 1)\delta_{ij}n_j/2\pi \qquad (3.23)$$

(δ_{ij} is the Krönecker delta) to

$$p_{RS}(i) = \varphi^{i*}(0,\mu = 1)n_i/R = \varphi^{i*}(0,\mu = 1)n_i/\sum_j \varphi^{j*}(0,\mu = 1)n_j . \quad (3.24)$$

In benchmark calculation problems 1 and 6 (Table V), in the examples of homogeneous and heterogeneous one-dimensional shields of iron and sodium, the authors of the present book used the method of direct substitution to analyze the sensitivity of the calculated results to the source geometry and the angular and energy distributions of the source radiation. The sensitivity of the results to the source geometry was analyzed by comparing the calculated results for a plane source (for a plane geometry of the shield) and for a point source (for a spherical geometry of the shield) while holding the other parameters of the problem fixed. Studies of the sensitivity of the data to the angular distribution of the source neutrons were carried out by comparing the results obtained in a plane geometry with plane neutron sources having the same spectrum but different angular distributions.

The sensitivity (R) to the energy distribution of the source neutrons has been studied by a generalized perturbation theory. Neutron sources with various energy compositions were considered in an analysis of the results (Table VI).

Just how influential the parameters of the source are depends on the particular characteristic of the radiation field which is being measured. Accordingly, several functionals (R) of the radiation field were selected for the analysis (Table V).

Figures 20–22 show the spatial distributions of various characteristics of the neutron field in plane and spherical, homogeneous and heterogeneous shields calculated by the ROZ-11 program for sources with various angular and energy distributions.

Comparison of the spatial distributions of the functionals found in the plane geometry with the corresponding distributions (multiplied by $4\pi x^2$) in the spherical geometry of the shield shows that these results differ in magnitude and form for the thicknesses of the iron and sodium shields studied. For

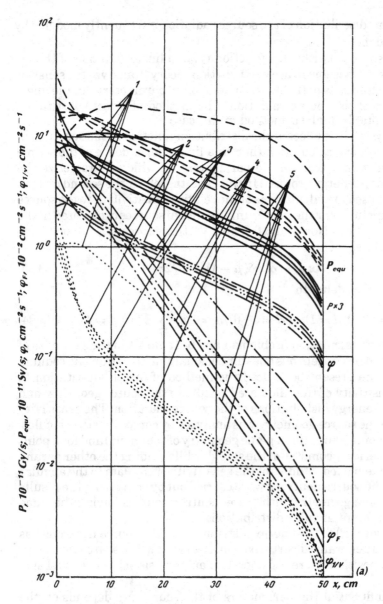

Figure 20. Spatial distributions in iron. ——, Tissue dose absorption rate P; ---, dose equivalent rate P_{equ}; ·-·-·-, total neutron flux density φ; -·-·-, fast-neutron flux density φ_F; ·····, readings of a $1/v$ detector, $\varphi_{1/v}$. 1, isotropic point source with a strength of 1 s^{-1} in spherical geometry. 5, isotropic plane source. 4, cosinusoidal plane source. 3, Cosine-cubed plane source ($\cos^3 \theta$). 2, unidirectional plane source in plane geometry (the source strength for lines 2, 3, 4, and 5 is $1\,cm^{-2}\,s^{-1}$). (a) The neutron energy distribution of the source is $f + 1/E$; (b) the energy distribution corresponds to an intermediate-neutron converter. The data for the isotropic point source have been multiplied by $4\pi x^2$.

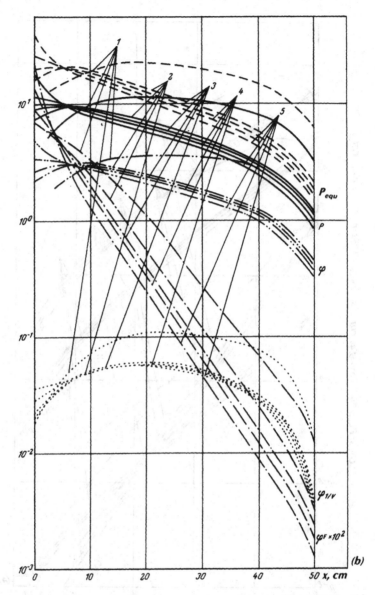

Figure 20. Cont'd.

the spherical geometry, the attenuation of the functionals is described by a function which is smoother than that for a plane geometry, and for the functionals considered it is not possible to find a simple description of the transformation from one geometry to the other. This circumstance limits the use of the neutron attenuation characteristics in shielding (e.g., the neutron relaxation lengths) found for a unidirectional plane source in calculations for a spherical shield, as is frequently done in practice.

Comparison of the results obtained in plane geometry for sources with various angular distributions indicates that the shape of the angular distributions of the characteristics of the radiation field in a shield with a thickness

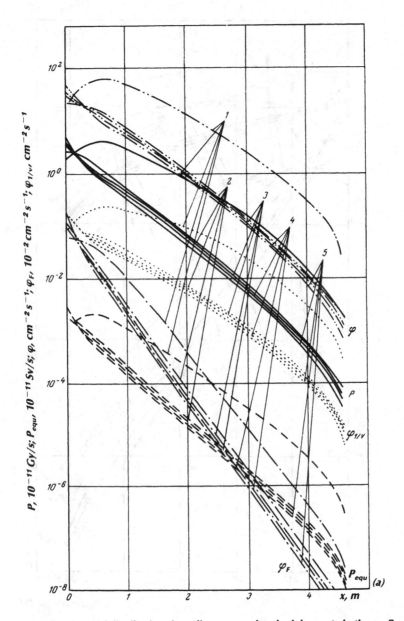

Figure 21. Spatial distributions in sodium. ——, absorbed dose rate in tissue, P; ---, dose equivalent rate P_{equ}; ·····, total neutron flux density φ; ----, fast-neutron flux density φ_F; ·····, readings of a 1/v detector, $\varphi_{1/v}$. 1, isotropic point source with a strength of 1 s^{-1} in spherical geometry. 5, isotropic plane source. 4, cosinusoidal plane source. 3, Cosine-cubed plane source (cos^3 θ). 2, unidirectional plane source in plane geometry (the source strength for lines 2, 3, 4, and 5 is 1 cm^{-2} · s^{-1}). (a) The neutron energy distribution of the source is $f + 1/E$; (b) the energy distribution corresponds to an intermediate-neutron converter. The data for the isotropic point source have been multiplied by $4\pi x^2$.

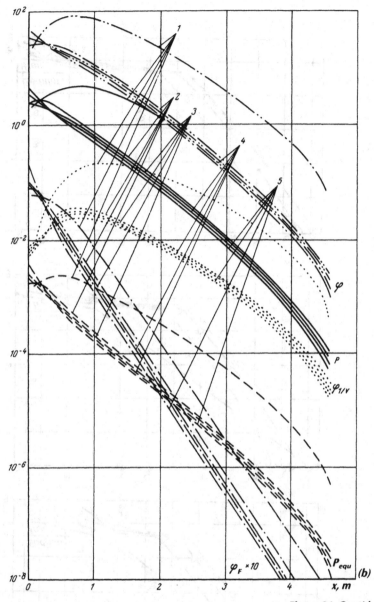

Figure 21. Cont'd.

$x \gtrsim 2L_F$ (L_F is the relaxation length of the fast neutrons) is independent of the angular distribution of the radiation from the source. As expected, the absolute values increase with increasing anisotropy of the angular distribution of the source neutrons.

The fact that the shape of the spatial distributions is independent of the angular distribution of the source radiation is explained on the basis that, except for the flux density of fast neutrons, the functionals of the radiation field considered here are determined in these media by moderated neutrons, as is shown by the results on their attenuation with increasing shield thick-

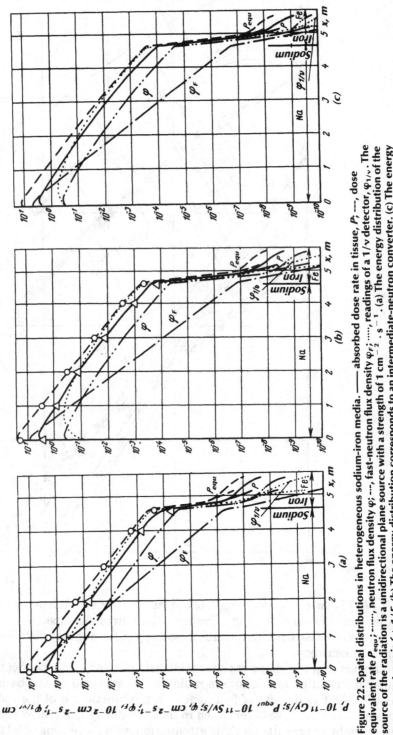

Figure 22. Spatial distributions in heterogeneous sodium–iron media. —— absorbed dose rate in tissue, P; ---, dose equivalent rate P_{equ}; ·····, neutron flux density φ; -··-, fast-neutron flux density φ_F; ···, readings of a 1/v detector, $\varphi_{1/v}$. The source of the radiation is a unidirectional plane source with a strength of 1 cm^{-2} · s^{-1}. (a) The energy distribution of the source neutrons is $f + 1/E$. (b) The energy distribution corresponds to an intermediate-neutron converter. (c) The energy distribution is that of a fast test reactor (FTR). The circles and triangles show data in a homogeneous sodium shield 4.6 m thick. The lower curves correspond to a shield with Na(4.6) + Fe(0.5).

Table XV. Coefficients for transforming from the data for a unidirectional source to the data for plane sources with other angular distributions of the radiation: isotropic (PI), $\sim \cos \theta$ (PC), and $\sim \cos^3 \theta$ (PC3).

Source	P	P_{equ}	φ	φ_F	φ_{ind}	$\varphi_{1/v}$
			Functional of radiation field			
		Iron ($x = 15$–50 cm); $f + 1/E$ initial spectrum				
PI	0.70 ± 0.01	0.68 ± 0.01	0.73 ± 0.01	0.41 ± 0.03	0.73 ± 0.01	0.75 ± 0.01
PC	0.80 ± 0.01	0.79 ± 0.01	0.83 ± 0.01	0.54 ± 0.03	0.83 ± 0.01	0.84 ± 0.02
PC3	0.88 ± 0.01	0.87 ± 0.01	0.90 ± 0.01	0.67 ± 0.03	0.90 ± 0.01	0.91 ± 0.01
		Iron ($x = 15$–50 cm); Initial spectrum corresponding to INC				
PI	0.76 ± 0.03	0.73 ± 0.02	0.80 ± 0.04	0.39 ± 0.03	0.80 ± 0.04	0.88 ± 0.04
PC	0.86 ± 0.02	0.83 ± 0.02	0.89 ± 0.03	0.52 ± 0.04	0.89 ± 0.03	0.95 ± 0.02
PC3	0.93 ± 0.01	0.91 ± 0.01	0.95 ± 0.02	0.66 ± 0.04	0.95 ± 0.02	0.99 ± 0.01
		Sodium ($x = 1$–4.6 m); $f + 1/E$ initial spectrum				
PI	0.66 ± 0.01	0.65 ± 0.01	0.67 ± 0.02	0.44 ± 0.05	0.67 ± 0.02	0.68 ± 0.01
PC	0.76 ± 0.01	0.75 ± 0.01	0.77 ± 0.01	0.56 ± 0.05	0.77 ± 0.01	0.77 ± 0.01
PC3	0.85 ± 0.01	0.84 ± 0.01	0.85 ± 0.01	0.69 ± 0.04	0.85 ± 0.01	0.86 ± 0.01
		Sodium ($x = 1$–4.6 m); Initial spectrum corresponding to INC				
PI	0.68 ± 0.01	0.67 ± 0.01	0.68 ± 0.01	0.48 ± 0.03	0.68 ± 0.01	0.69 ± 0.01
PC	0.77 ± 0.01	0.77 ± 0.01	0.77 ± 0.01	0.60 ± 0.03	0.77 ± 0.01	0.78 ± 0.01
PC3	0.86 ± 0.01	0.86 ± 0.01	0.86 ± 0.01	0.72 ± 0.03	0.86 ± 0.01	0.87 ± 0.01

ness and by the fact that the total neutron flux density is determined entirely by neutrons with energies below 1.4 MeV. We might thus say that a sort of quasisource of moderated neutrons with an equilibrium spectrum is formed at a shielding thickness $x \approx 2L_F$. The angular distribution of the radiation from this source is essentially independent of the angular distribution of the radiation from the source because of the particular way in which the neutrons are moderated in iron and sodium; only the strength of the quasisource changes.

Table XV shows coefficients for transforming from the data obtained for a unidirectional plane source to data for sources with other angular distributions for homogeneous iron and sodium shields; these transformation coefficients are averaged over the shield thickness. These coefficients essentially tell us the sensitivity of the calculated results to the angular distribution of the source neutrons. With the exception of the flux density of fast neutrons, the effect of the angular distribution of the source neutrons does not exceed 45%–50% for these homogeneous shields, regardless of the original spectrum of the source. For the particular case of the flux density of fast neutrons, this difference can reach a factor of 2 or 3. The designer of a shield usually has more accurate information on the angular distribution of the source neutrons, so that the uncertainties due to the angular distribution of the source neutrons in the design of an actual shield will be lower.

The nature of the neutron attenuation in the first sodium layer in heterogeneous sodium-iron shields and the dependence of the field characteristics

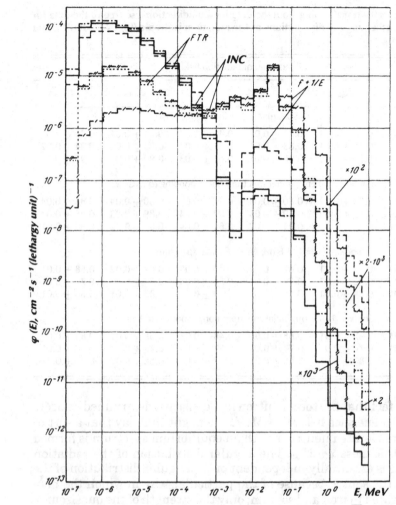

Figure 23. Energy flux densities of neutrons in Na at a depth of 4.5 m (- -) and in Fe at a depth of 4.9 m (....‖....) in a composite shield consisting of Na(4.6 m) + Fe(0.5 m) according to calculations for unidirectional plane sources with a strength of 1 cm^{-1}·s^{-1} with an $f + 1/E$ neutron energy distribution (—/ —), with a neutron energy distribution corresponding to an intermediate-neutron converter (—‖—) and an FTR neutron energy distribution (.......).

on the angular distribution of the source neutrons are similar to those which are observed in a homogeneous sodium shield. The attenuation of neutrons in iron is substantially different. The reason is that the energy distribution of the neutrons beyond the sodium layer becomes rich in low-energy neutrons, regardless of the neutron spectrum of the source. These low-energy neutrons are then effectively absorbed by the iron through radiative capture. The effective attenuation of the functionals of the neutron field occurs over the first 10 cm; as the original source spectrum becomes softer the spectrum of neutrons beyond the sodium layer also become softer, and the attenuation of the functionals in the iron layer becomes more effective.

Table XVI. Coefficients for transforming from the data for a unidirectional plane source to data for an isotropic source for a shield with the composition Na(4.6 m) + Fe (0.9 m).

Shield thickness	Functional of radiation field					
	P	P_{equ}	φ	φ_F	φ_{ind}	$\varphi_{1/v}$
f + 1/E initial spectrum						
Homogeneous Na	0.66	0.65	0.67	0.44	0.67	0.68
$x = 4.6$ m	0.64	0.64	0.64	0.37	0.64	0.66
$x = 4.75$ m	0.59	0.57	0.6	0.36	0.6	0.63
$x = 5.2$ m	0.48	0.48	0.48	0.36	0.48	0.48
$x = 5.5$ m	0.48	0.48	0.48	0.36	0.48	0.48
Homogeneous Fe	0.70	0.68	0.73	0.41	0.73	0.75
Initial spectrum correspond to an intermediate-neutron converter						
Homogeneous Na	0.68	0.67	0.68	0.48	0.68	0.69
$x = 4.6$ m	0.66	0.66	0.66	0.38	0.66	0.66
$x = 4.75$ m	0.65	0.65	0.65	0.38	0.65	0.66
$x = 5.2$ m	0.59	0.58	0.58	0.33	0.58	0.58
$x = 5.5$ m	0.58	0.58	0.58	—	0.58	0.58
Homogeneous Fe	0.76	0.73	0.80	0.39	0.80	0.88
FTR initial spectrum						
$x = 4.6$ m	0.66	0.66	0.67	0.38	0.67	0.67
$x = 4.75$ m	0.65	0.65	0.65	0.38	0.65	0.65
$x = 5.2$ m	0.57	0.57	0.57	0.37	0.57	0.57
$x = 5.5$ m	0.57	0.57	0.57	—	0.57	0.57

Confirmation of this fact can be seen in the energy flux density of the neutrons in a heterogeneous sodium-iron shield (Fig. 23). The increase in the role of the higher-energy neutrons in the iron has the consequence that the effect of the angular distribution of the source neutrons on the result of the calculations becomes stronger than for a homogeneous shield.

Table XVI shows coefficients for transforming from the data from a unidirectional neutron source to data for an isotropic source for various original spectra. We see that the maximum error in most of the functionals due to the error in the description of the angular distribution of the source radiation increases to 100%; i.e., it is about twice that for a homogeneous shield. Beginning at an iron thickness of 30 cm, i.e., after an equilibrium spectrum has been established, this error is independent of the thickness, as in the case of a homogeneous shield.

The sensitivity of the calculated results to the energy distribution of the source neutrons has been determined from data from energy channel theory. These calculations were carried out by perturbation theory in the ZAKAT program for four initial spectra: a fission spectrum (f), an $f + 1/E$ spectrum, the spectrum corresponding to an intermediate-neutron converter, and an FTR spectrum, for the set of neutron detectors described above, positioned beyond the shields.

Since this relative sensitivity of the calculated results depends relatively weakly on the angular distribution of the radiation from a plane source (Fig.

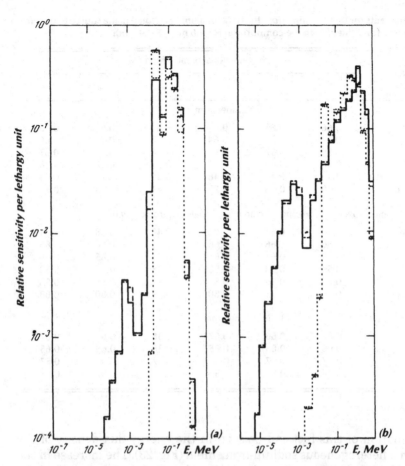

Figure 24. Relative sensitivity of the equivalent dose rate P_{equ} to the neutron energy distribution from an infinite plane isotropic source (---,-/-) and an infinite plane unidirectional source (...,——) with initial spectra corresponding to (a) an intermediate-neutron converter and (b) $f + 1/E$. (...,---/---), in Fe; (——,---), in Na.

24), we will consider below the relative sensitivities of the results to the energy distribution of the source neutrons for an infinite, plane, unidirectional source.

Information on the effect of the energy distribution of the source neutrons on the result calculated for an arbitrary source spectrum can be found if we know the importance function determined at the position of the source. That function is proportional to the contribution to the functional R from one particle in the given energy group i emitted by the neutron source.

Figures 25–27 illustrate the situation with importance functions of the neutron source calculated by the ROZ-11 program. The energy dependence of the importance function for the shield configurations studied shows that high-energy neutrons from the source play a significant role in the formation of any functional. It follows that if the initial spectrum of the neutron source is rich in fast neutrons (an example would be the spectrum of leakage neu-

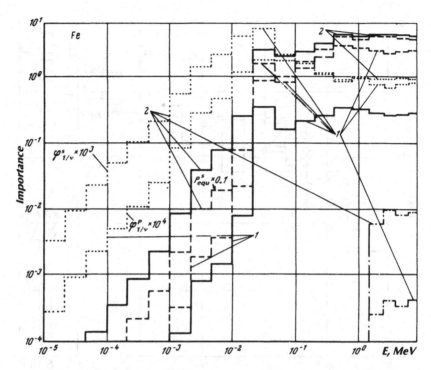

Figure 25. Importance function of neutrons at the entrance to a shield. ——,
absorbed dose rate in tissue, P; ---, dose equivalent rate P_{equ}; (—— · ——), fast-
neutron flux density φ_F; (.....), readings of a 1/v detector, $\varphi_{1/v}$, behind a 50-cm-thick
Fe shield. 1, plane geometry; 2, spherical geometry.

trons from a fusion reactor), then essentially all the functionals of the neutron field in an iron-sodium shield will be determined by the high-energy
neutrons of the source. For a "white" spectrum of source neutrons, the calculated functionals in iron and sodium are dominated by neutrons with energies $E > 0.05$ MeV. In this range, their contributions to the energy dependence
depend only weakly on the energy in a case of iron, while they increase quite
rapidly with increasing energy in sodium. In heterogeneous sodium-iron configurations the high-energy neutrons will be even more important than in
homogeneous shields. Source neutrons with energies $E > 0.5$ MeV are dominant in this case. It should be noted that the energy dependence of the importance function of the source neutrons retains its shape when the thickness of
the second (iron) layer is changed. The absolute values of the importance
function for the absorbed and dose equivalent rates behind a composite shield
consisting of Na(4.6 m) + Fe(0.5 m) can be found from the data in Fig. 27 by
multiplying by 2.5. In reactor problems, the spectra of the leakage from the
core are significantly enriched in intermediate-energy neutrons, and in this
connection the sensitivity of the calculated results depends on the ratio of the
contributions of fast and intermediate-energy neutrons in the initial spectrum.

The situation can be seen quite clearly in the energy dependence of the
relative sensitivity of various functionals to the parameters of the functions

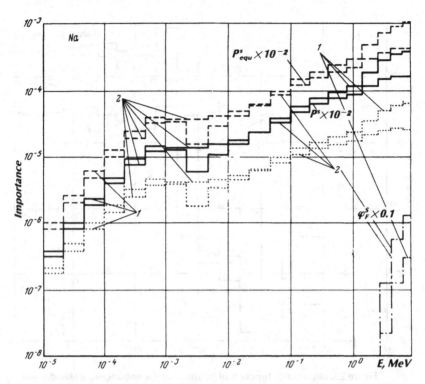

Figure 26. Neutron importance functions at the entrance to a shield. ——, absorbed dose rate in tissue, P; ---, dose equivalent rate P_{equ}; (—— · ——), fast-neutron flux density φ_F; ....., readings of a 1/v detector, $\varphi_{1/v}$, behind a 4.6-m-thick Na shield. 1, plane geometry; 2, spherical geometry.

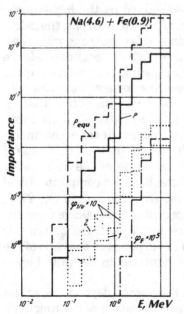

Figure 27. Neutron importance functions at the entrance to a shield. ——, absorbed dose rate in tissue, P; ---, dose equivalent rate P_{equ}: —— · ——, fast-neutron flux density φ_F; ....., readings of a 1/v detector, $\varphi_{1/v}$, behind a shield consisting of Na(4.6) + Fe(0.9). 1, plane geometry; 2, spherical geometry.

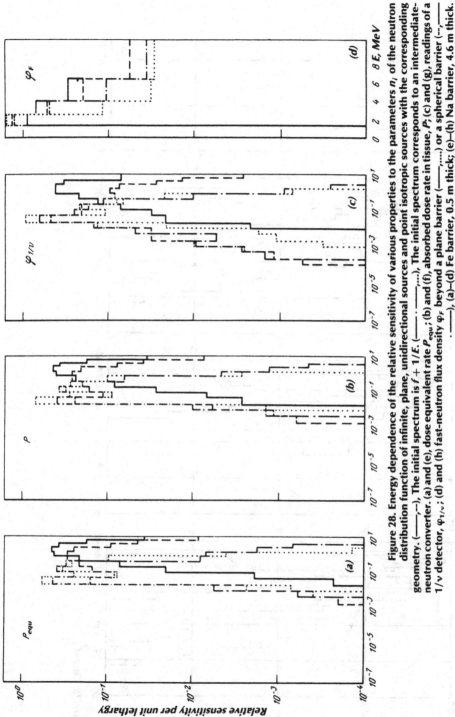

Figure 28. Energy dependence of the relative sensitivity of various properties to the parameters n_i of the neutron distribution function of infinite, plane, unidirectional sources and point isotropic sources with the corresponding geometry. (———,——), The initial spectrum is $f + 1/E$; (———. ———,....), The initial spectrum corresponds to an intermediate-neutron converter. (a) and (e), dose equivalent rate P_{equ}; (b) and (f), absorbed dose rate in tissue, P; (c) and (g), readings of a $1/v$ detector, $\varphi_{1/v}$; (d) and (h) fast-neutron flux density φ_F beyond a plane barrier (———,....) or a spherical barrier (—,———), (a)–(d) Fe barrier, 0.5 m thick; (e)–(h) Na barrier, 4.6 m thick.

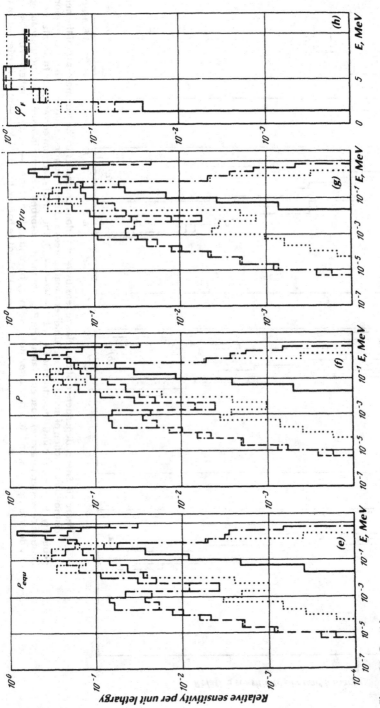

Figure 28. Cont'd.

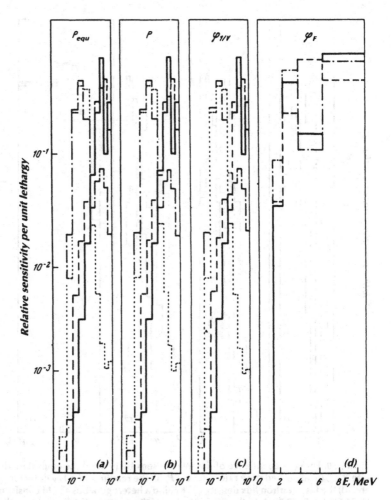

Figure 29. Energy dependence of the relative sensitivity of various properties to the parameters n_i of the neutron distribution function of infinite plane unidirectional sources. ——, initial spectrum is f. –·–, initial spectrum is $f + 1/E$. ······, initial spectrum corresponds to an intermediate-neutron converter. ----, initial spectrum corresponds to an FTR. (a) Equivalent dose rate P_{equ}; (b) absorbed dose rate in tissue, P; (c) readings of a $1/v$ detector, $\varphi_{1/v}$; (d) fast-neutron flux density φ_F, behind a heterogeneous shield consisting of Na (4.6 m) + Fe (0.5 m).

which specify the neutron distributions of reactor sources with various neutron compositions, shown in Figs. 28–30 and Tables XVII and XVIII. These results are known to characterize the relative change $(\delta R/R)$ in a functional upon a relative change $(\delta n_i/n_i)$ in the number (n_i) of neutrons in energy group i. They can be used to estimate the error in a result which is due to the errors in the specification of source group i.

Analysis of the energy dependence of the relative sensitivity to the neutron distribution function of the source reveals certain governing energy groups of the source neutrons, on which the most stringent requirements are imposed for the specification of the energy spectrum. Tables XVII and XVIII show the neutron groups in the spectrum of a unidirectional plane source

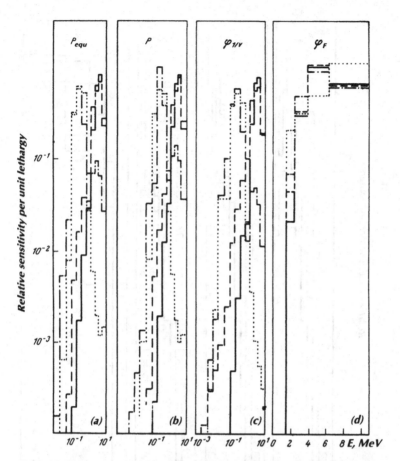

Figure 30. Energy dependence of the relative sensitivity of (a) the equivalent dose rate P_{equ}, (b) absorbed dose rate in tissue, P, (c) the readings of a $1/v$ detector, $\varphi_{1/v}$, and (d) the fast-neutron flux density φ_F, behind a heterogeneous shield consisting of Na (4.6 m) + Fe (0.9 m), to the parameters n_i of the neutron distribution function of an infinite, plane, unidirectional source. The initial source spectrum is ———, f; ---, $f + 1/E$; ....., spectrum corresponding to an intermediate-neutron converter; ·-·-·, FTR spectrum.

which make the governing contribution to the relative sensitivity of the calculated results for the benchmark calculation problems. The basic errors in the calculated results may thus result from source neutrons with energies from 21.5 keV to 4.0 MeV and from 21.5 to 800 keV in Fe for the $f + 1/E$ spectrum and for the spectrum corresponding to an intermediate-neutron converter, respectively; the corresponding intervals in Na are from 0.465 keV to 10.5 MeV and from 46.5 to 800 keV for the respective spectra; and the corresponding interval for heterogeneous shields of Na and Fe is from 0.2 to 10.5 MeV.

The major role played by high-energy neutrons from the source in determining the readings of any detector of R, even for such relatively soft spectra as the $f + 1/E$ spectrum and that corresponding to an intermediate-neutron converter, points out the need for a precise specification of the high-energy

Table XVII. Indices of the neutron groups of a unidirectional plane source which make the governing contribution to the calculated results behind Fe and Na.

Functional R	Contributions to functional R, %							
	for Fe (0.5)				for Na (4.6)			
	20–30	50–60	90–93	99–99.5	20–30	50–60	90–93	99–99.5
	$f + 1/E$ initial spectrum							
P	10	—	4–10	2–10	4	3–6	2–11	1–16
P_{equ}	6	5–7	4–10	3–10	4	3–6	2–11	1–16
$\varphi_{1/v}$	10	9–11	4–11	3–13	4	4–7	1–12	1–17
φ_F	—	4	2–4	—	—	2	1–3	—
	Initial spectrum corresponding to an intermediate-neutron converter							
P	—	10	7–10	6–10	8	7–8	7–10	6–10
P_{equ}	—	10	7–10	6–10	8	7–8	7–10	6–10
$\varphi_{1/v}$	—	10	8–10	7–10	8	7–8	7–10	6–10
φ_F	—	—	4	—	—	1	—	—

Table XVIII. Indices of the neutron groups of a undirectional plane source which make the governing contribution to the calculated results behind heterogeneous Na + Fe shields.

Functional R	Contribution to R, %							
	for Na (4.6) + Fe (0.5)				for Na (4.6) + Fe (0.9)			
	30–40	60–70	90–93	99–99.5	30–40	60–70	90–93	99–99.5
	Initial spectrum corresponding to an intermediate-neutron converter							
P	7	6–7	6–8	5–8	7	6–7	—	5–8
P_{equ}	7	6–7	6–8	5–8	7	6–7	—	5–8
$\varphi_{1/v}$	7	6–7	6–8	5–10	7	6–7	—	5–8
φ_F	—	1	1–3	—	—	1	—	—
	$f + 1/E$ initial spectrum							
P	2	2–3	1–4	1–6	2	2–4	1–4	1–8
P_{equ}	2	2–3	1–4	1–6	2	2–4	1–4	1–8
$\varphi_{1/v}$	2	2–4	1–5	1–7	2	2–4	1–4	1–8
φ_F	—	2	1–3	—	—	1–2	1–3	—
	FTR initial spectrum							
P	7	7–8	4–8	2–9	7	7–8	4–8	2–9
P_{equ}	7	7–8	4–8	2–9	7	7–8	4–8	2–9

tails of these spectra. This need is not obvious at first glance. The results on the importance function for source neutrons shown in Figs. 25 and 26 for homogeneous spherical shields of iron and sodium show that in spherical geometry the source neutrons with lower energies become more important than in plane geometry. The energy range of the source neutrons which basically govern the relative sensitivity of the result of a calculation thus becomes broader for the spherical geometry, extending toward lower energies.

The data obtained on the relative sensitivity (P_{RS}) of the results of the calculations of various functionals to the neutron energy distribution (n_i) of sources with specific initial spectra can be used to determine the corresponding relative sensitivities P'_{RS} for the given shield configurations for neutron sources with an arbitrary spectrum n'_i. We obviously have

$$p'_{RS}(i) = p_{RS}(i) \frac{n'_i}{n_i} \frac{R}{R'}, \tag{3.25}$$

where R and R' are the results calculated for the neutron sources with different energy compositions. Since in many practical problems, especially in the first stage of a design project, it is important to estimate the energy range in the source emission spectrum which is basically responsible for the error in the functional being determined—the goal at this point is not to determine the absolute value of the relative sensitivity—we can omit the factor R/R' in (3.25) (this factor does not depend on the neutron energy) and use the data in Figs. 28–30 to determine the energy dependence of the relative sensitivity p'_{RS} for a source with an arbitrary neutron spectrum.

Let us examine the sensitivity of the calculated results to the parameters of the detector response function. Among the characteristics of a detector which contribute errors to the result we can single out the following: (1) the overall detection efficiency, (2) the energy dependence of the detection efficiency, (3) the angular dependence of the detection efficiency, and (4) the detector geometry.

For an isotropic point detector with a given detection efficiency, the only parameter which would affect the error in the calculated result is the energy dependence of the conversion coefficients $\delta_R(E)$ for converting from a unit fluence of neutrons with a given energy to the particular functional (R) of the radiation field which is being calculated (see, for example, Table I).

In the group approximation of the energy variable, there is an exact linear relationship between the detector response function $\delta_R(E)$ and the parameters characterizing the efficiency of the detection of the neutrons of group i. In an analysis of the sensitivity of the calculated results to the energy dependence of the detection efficiency or of the conversion coefficients $\delta_R(E)$, we can thus use the first term in expression (1.29). For an isotropic point detector at the point x_0, beyond a shield of a certain thickness, which is detecting neutrons of group i with an efficiency δ_R^i,

$$q^i(x,\mu) = \delta(x - x_0)\delta_{ij}\delta_R^j/4\pi, \tag{3.26}$$

where δ_{ij} is the Krönecker delta; this expression becomes

$$p_{Rq}(i) = \varphi^i(x_0)\delta_R^i/R. \tag{3.27}$$

It can be seen from (3.27) that the relative sensitivity of the calculated result to the detector response function depends on the energy distribution of the neutrons beyond the shield, $\varphi^i(x_0)$, and on the energy dependence of δ_R^i.

For benchmark calculation problems 1 and 6 (Table V), the relative sensitivities of various functionals behind homogeneous and heterogeneous shields of Na and Fe were calculated with the ROZ-11 + ZAKAT + ARAMAKO-2F software for unidirectional plane sources with various energy distributions of

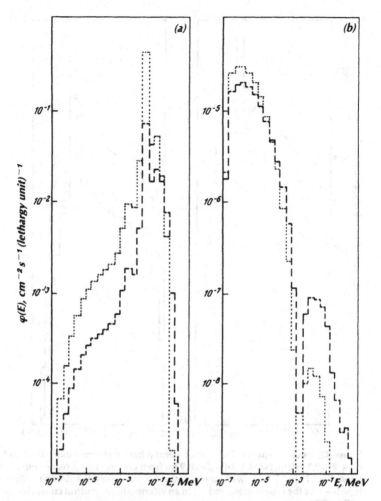

Figure 31. Neutron energy distributions behind homogeneous barriers (a) of Fe, with a thickness of 0.5 m, and (b) of Na, with a thickness of 4.6 m, from a unidirectional plane source with a strength of 1 cm$^{-2} \cdot$ s^{-1} with an emitted-neutron energy distribution of the $f + 1/E$ type (----) and of the type corresponding to an intermediate-neutron converter (....).

the emitted neutrons. Since the relative sensitivity of the results depends on the neutron energy distribution beyond the barriers and the particular values of δ_R^i which are used, these energy neutron distributions calculated from the ROZ-11 program and the conversion coefficients used are of interest for interpreting the results. Figures 31 and 32 show, in a 26-energy-group representation, the energy distributions of the neutrons behind homogeneous and heterogeneous barriers of Na and Fe for unidirectional plane sources with neutron energy distributions of the $f + 1/E$ type, the type corresponding to an intermediate-neutron converter, and the FTR type.

As mentioned earlier, the neutron energy distribution behind an iron shield has a significant contribution of intermediate neutrons, especially in the energy interval 21.5–46.5 keV, corresponding to a minimum in the iron

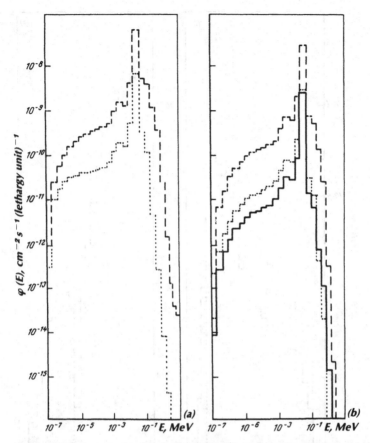

Figure 32. Neutron energy distribution behind heterogeneous barriers of (a) Na (4.6 m) + Fe (0.5 m), (b) Na (4.6 m) + Fe (0.9 m) from unidirectional plane sources with a strength of 1 cm^{-2}·s^{-1} with an emitted-neutron energy distribution of the $f + 1/E$ type (----), of the type corresponding to an intermediate-neutron converter (····), and of the FTR type (——).

cross section at 25 keV. The contribution of the neutrons of this group to the total neutron flux density behind the shield is 50% for the $f + 1/E$ source and 73% for the source corresponding to an intermediate-neutron converter.

The energy distribution of the neutrons behind a sodium shield exhibits a buildup of slow neutrons with energies from 10^{-6} to 10^{-4} MeV. The contribution of neutrons with energies from 0.465 to 21.5 eV to the total neutron flux density behind a sodium shield 4.6 m thick is 80% for the $f + 1/E$ source and 85% for the source corresponding to an intermediate-neutron converter.

The neutron spectrum which forms behind heterogeneous Na + Fe shields approaches the neutron spectrum in pure Fe as a result of the intense absorption in the Fe of the slow neutrons built up in the Na. There is an even more prominent peak at 25 keV. Behind heterogeneous shields, the contribution of this neutron group (21.5–46.5 keV) to the total flux density is 73% for the composition Na(4.6 m) + Fe(0.5 m) and 82% for Na(4.6 m) + Fe(0.9 m) for an $f + 1/E$ source spectrum; the corresponding figures for a source spectrum

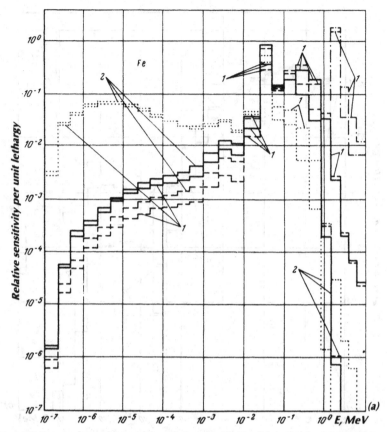

Figure 33. Energy dependence of the relative sensitivity of the absorbed dose rate in tissue, P (——), of the dose equivalent rate P_{equ} (----), of the readings of a $1/v$ detector, $\varphi_{1/v}$ (.....), and of the fast-neutron flux density φ_F (——·——), behind plane barriers of (a) Fe and (b) Na to the detector response function for unidirectional plane sources with an $f + 1/E$ spectrum (1) and the spectrum corresponding to an intermediate-neutron converter (2).

corresponding to an intermediate-neutron converter are 78% and 84%; and those corresponding to an FTR source spectrum are 70% and 84%.

The energy dependence of the detector response function also changes substantially when we go from one detector to another. Figures 33–35 show the energy dependence of the relative sensitivity of these functionals R to the parameters δ_R^i of the detector response function for various shielding compositions and for various initial source spectra. From these results we can determine those energy groups in which we need the most accurate specification of the energy characteristics of the detector in order to reduce the errors in the calculated results. Tables XIX and XX show those energy groups in the 26-group representation for the detector response function which make the primary contribution to the calculated results behind homogeneous and heterogeneous shields of Na and Fe. As expected, behind the Fe barrier, regardless of the source spectrum and regardless of the particular detector used, the basic contribution to the calculated results and thus to its errors comes from

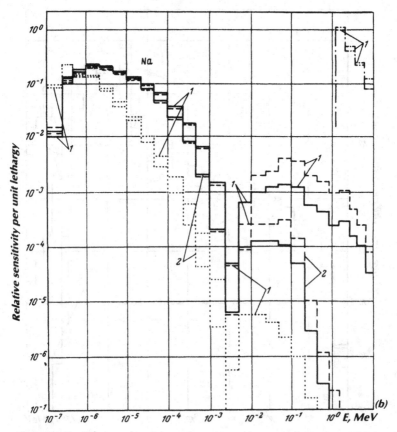

Figure 33. Cont'd.

the neutrons with energies in the range 21.6–46.5 keV. For a source spectrum corresponding to an intermediate-neutron converter, the contribution of this group exceeds 50%. More than 90% of the contribution to the response of the detector comes from the rather narrow energy interval 0.02–0.8 MeV for the absorbed dose rate and the equivalent dose rate, 2.15×10^{-6}–0.4 MeV for $\varphi_{1/v}$, and 1.4–2.5 MeV for the fast-neutron flux density. These energy intervals are essentially independent of the radiation energy spectrum of the source.

Behind the Na barriers, the basic contribution again comes from neutrons in a rather narrow energy interval, but in this case it is an interval of slow neutrons. More than half of the contribution to the detector response comes from neutrons with energies in the interval 0.465–10 eV, regardless of the functional being calculated and regardless of the source spectrum. More than 90% of the contribution comes from neutrons with energies between 0.025 and 100 eV for the absorbed dose rate and the dose equivalent rate, between 0.025 and 4.65 eV for $\varphi_{1/v}$, and between 1.4 and 6.5 MeV for the fast-neutron flux density. For the heterogeneous Na + Fe shields, the behavior is similar to that observed in the homogeneous iron shield. The role of the interference minimum in the neutron-iron interaction cross section at 25 keV increases by a factor of about 2, however, and while its relative contribution in a homogeneous shield is no more than 30% it approaches 70% in a hetero-

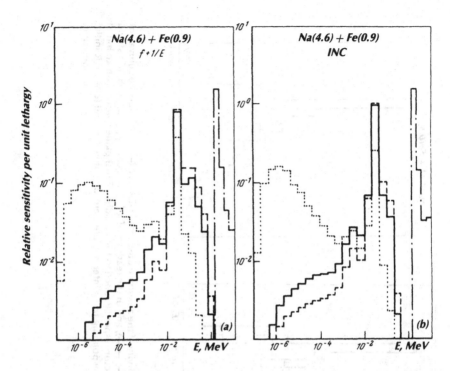

Figure 34. Energy dependence of the relative sensitivity of the absorbed dose rate in tissue, P (———), of the equivalent dose rate P_{equ} (----), of the fast-neutron flux density φ_F (——·——), and of the readings of a $1/v$ detector, $\varphi_{1/v}$ (...), behind a shield of Na (4.6 m) + Fe (0.5 m) to the detector response function for unidirectional plane sources with (a) an $f + 1/E$ spectrum and (b) a spectrum corresponding to an intermediate-neutron converter.

geneous shield and increases with increasing thickness of the second iron layer.

There is yet another feature of the effect of the neutron energy distribution of the source on the results of a sensitivity analysis to be noted. This point was mentioned in Ref. 3. In an analysis of the sensitivity of calculated results to the neutron-matter interaction cross sections, the initial spectrum of the source is important, since a relatively slight change in the source spectrum leads to significant changes in the picture of the sensitivity of the result to the interaction cross sections. For example, we consider calculation problem 3 (Table V), involving a comparison of the data on the relative sensitivity of the readings of a detector with a constant detection efficiency over the energy range 40–350 keV, positioned behind a sodium shield with a thickness of 5 m, to the total neutron-sodium interaction cross section for two initial source spectra. It was shown in this problem (as was mentioned in Sec. 1.4) that for a softer source spectrum the minimum in the Na cross section at 300 keV determines the relative sensitivity almost entirely (61.4%), while the corresponding figure for a hard spectrum is 2.3%.

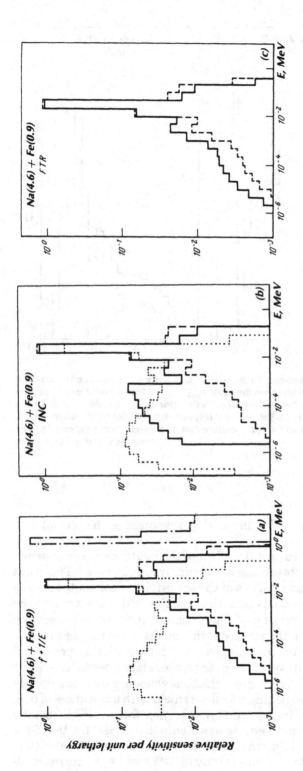

Figure 35. Energy dependence of the relative sensitivity of the absorbed dose rate in tissue, P (——), of the dose equivalent rate P_{equ} (——), of the fast-neutron flux density (— . ——), and of the readings of a '1/v detector, $\varphi_{1/v}$ (...), behind a shield consisting of Na (4.6 m) + Fe (0.9 m) to the detector response function for a unidirectional plane source with an initial spectrum (a) $f + 1/E$, (b) an initial spectrum corresponding to an intermediate-neutron converter, and (c) an FTR initial spectrum.

Table XIX. Indices of the energy groups of a 26-group representation of the detector response function which make a governing contribution to the results of a calculation of R behind shields of Fe and Na.

	Contribution to functional R, %							
	Fe				Na			
Functional R	20–30	50–60	90–93	99–99.5	20–30	50–60	90–93	99–99.5
	$f + 1/E$ initial spectrum							
P	10	—	6–10	5–11	22–23	21–24	18–26	14–26
P_{equ}	10	7–10	6–10	5–11	22–23	21–24	18–26	14–26
$\varphi_{1/v}$	10	7–14	7–23	7–26	—	25–26	22–26	20–26
φ_F	—	—	4	2–4	—	4	2–4	—
	Initial spectrum corresponding to an intermediate-neutron converter							
P	—	10	8–10	7–10	22–23	21–24	19–26	17–26
P_{equ}	—	10	8–10	7–10	22–23	21–24	19–26	17–26
$\varphi_{1/v}$	—	10	8–22	8–25	—	25–26	22–26	20–26
φ_F	—		4	3–4	—	4	2–4	—

Table XX. Indices of the energy groups of a 26-group representation of the detector response function which make the governing contribution to the results of a calculation of R behind heterogeneous shields of Na and Fe.

	Contribution to functional R, %							
	Na(4.6) + Fe(0.5)				Na(4.6) + Fe(0.9)			
Functional R	30–40	60–70	90–93	99–99.5	30–40	60–70	90–93	99–99.5
	$f + 1/E$ initial spectrum							
P	—	10	7–11	5–20	—	10	8–11	7–17
P_{equ}	—	10	7–11	5–20	—	10	8–11	7–17
$\varphi_{1/v}$	10	9–20	9–24	8–26	10	10–19	10–23	10–26
φ_F	—	—	4	2–4	—	—	4	—
	Initial spectrum corresponding to an intermediate-neutron converter							
P	—	10	8–11	7–22	—	10	10–11	8–17
P_{equ}	—	10	8–11	7–22	—	10	10–11	8–17
$\varphi_{1/v}$	10,23,24	—	10,19,26	8–26	10	10–19	10–23	10–26
φ_F	—	—	4	2–4	—	—	4	—
	FTR initial spectrum							
P	—	—	10–11	8–17	—	—	10–11	8–17
P_{equ}	—	—	10–11	8–17	—	—	10–11	8–17

Table XXI. Calculated flux densities of fast neutrons with energies $E > 1.4$ MeV, φ_F; the tissue dose absorbed dose rate P; the dose equivalent rate P_{equ}; and the readings of a $1/v$ detector, $\varphi_{1/v}$, behind one-dimensional plane heterogeneous shields of Na and Fe, normalized to a plane-source strength of 1 cm^{-2} s^{-1}.

| Radiation source | | Functional R | | |
	φ_F, cm^{-2} s^{-1}	P, Gy/s	P_{equ}, Sv/s	$\varphi_{1/v}$, arb. units
		Na(4.6 m) + Fe(0.5 m)		
$f + 1/E$	9.22–13[a]	4.53–17	2.49–16	7.27–5
INC	2.96–16	4.41–18	2.20–17	1.21–5
		Na(4.6 m) + Fe(0.9 m)		
$f + 1/E$	1.37–15	1.86–17	9.36–17	2.75–5
INC	3.73–19	1.90–18	9.26–18	2.94–6

[a] Read $a - b$ as $a \times 10^{-b}$.

It follows from Eq. (1.29) that the product $p_{Rn} R$ is a bilinear functional of φ^* and φ and therefore of the source specification function S and the detector response function q. The relative sensitivities $\{\{p_{Rn}^{km}\}_k\}_m$ of the functionals R^{km} (with detector response functions $\{q_k\}_k$) to the parameter X_n for the sources $\{S_m\}_m$ can therefore be used to find the relative sensitivity P_{Rn} of the functional R (with the detector response function $q = \sum_k \alpha_k q_k$) to the parameter X_n for the source $S = \sum_m \beta_m S_m$ with the help of the following expression:

$$p_{Rn} = \frac{1}{R} \sum_k \sum_m \alpha_k \beta_m R^{km} p_{Rn}^{km}. \qquad (3.28)$$

This expression is exact. It can be used to generalize the conclusions derived in a sensitivity study with one set of sources and detectors to other problems, in which the sources and detectors are related in a linear way to the sources and detectors, respectively, which have been studied.

Analysis of the latter expression leads to the following conclusion: Studies of the sensitivity of shielding calculations to the input parameters for some particular set of sources and detectors can be optimized by choosing for the studies only those sources and detectors which cannot be expressed in a linear way in terms of others.

To conclude this section we show in Tables XXI and XXII the absolute values of the calculated functionals behind the shields studied for various parameters of the source and detector. These results provide indirect information on the joint effects of the characteristics of the source and the detector on the calculated result.

Table XXII. Flux density of fast neutrons with energies $E \geqslant 1.4$ MeV, φ_F; the tissue absorbed dose rate P; the dose equivalent rate P_{equ}; and the readings of a $1/v$ detector, $\varphi_{1/v}$, behind plane shields of Na and Fe, normalized to a plane-source strength of 1 cm^{-2} s^{-1} for two source energy spectra, $f + 1/E$ and the spectrum corresponding to an intermediate-neutron converter, and four angular distributions (plane isotropic, plane cosinusoidal, cosine-cubed plane, and plane unidirectional).

Source energy spectrum	$f + 1/E$				INC			
Functional	PI	PC	PC3	PU	PI	PC	PC3	PU
				Iron				
φ_F, cm^{-2} s^{-1}	1.46– 5[a]	1.93– 5	2.95– 5	3.86– 5	1.38– 8	1.86– 8	2.40– 8	3.92– 8
P, Gy/s	7.86–13	9.03–13	1.00–10	1.14–12	2.58–12	2.98–12	3.28–12	3.62–12
P_{equ}, Sv/s	5.67–12	6.53–12	7.26–12	8.36–12	1.49–11	1.72–11	1.91–11	2.13–11
$\varphi_{1/v}$, arb. units	6.22– 5	7.00– 5	7.59– 5	8.26– 5	2.98– 4	3.35– 4	3.58– 4	3.76– 4
				Sodium				
φ_F, cm^{-2} s^{-1}	1.13– 9	1.51– 9	1.93– 9	3.06– 9	3.77–13	5.04–13	6.43–13	9.50–13
P, Gy/s	3.00–16	3.47–16	3.91–16	4.67–16	4.20–16	4.81–16	5.37–16	6.32–16
P_{equ}, Sv/s	7.68–16	8.8 –16	1.00–15	1.20–15	1.07–15	1.22–15	1.37–15	1.61–15
$\varphi_{1/v}$, arb. units	6.57– 6	7.56– 6	8.46– 6	9.98– 6	1.02– 5	1.17– 5	1.36– 5	1.52– 5

[a] Read $a - b$ as $a \times 10^{-b}$.

3. Estimates of the error in the calculated results and study of the sensitivity of the results to the neutron interaction cross sections for homogeneous and heterogeneous media

Errors in the calculated results and sensitivity of the results to inaccuracies in the interaction cross sections are always a major concern of researchers. We examine these questions in the present section. Only in the simplest of cases, in which the solution of the problem can be found by an analytic or semiempirical method, is it possible to find an analytic expression for the relative sensitivity in terms of the input parameters.

In other cases, such studies must generally be carried out on the basis of numerical solutions of the transport equation, a generalized perturbation theory, or the direct-substitution method.

Errors in the analytic methods

An analytic solution of radiation-transport problems can be found for homogeneous media under very restrictive assumptions regarding the energy dependence of the cross sections, frequently without consideration of the anisotropy of the neutron scattering. Despite the fundamental limitations of these methods, they have played and continue to play a useful role in the testing of numerical calculation methods.

Among the analytic methods for shielding calculations, we can identify essentially only one which is of definite practical significance. This method is the asymptotic solution of the radiation transport equation, found by the Fourier method or the Case method. A solution of this sort is convenient for an analytic study of the qualitative behavior of the relative sensitivity to the parameters of a problem.

The asymptotic value of the neutron flux density from a unidirectional plane source, $\delta(\mu - \mu_0)$, of unit strength in an infinite, homogeneous, isotropic scattering medium in the single-velocity problem is[20]

$$\varphi^{\,ac}(x;\mu_0) = \left[\nu \middle/ \left(\frac{\Sigma_S \Sigma_t}{\Sigma_t^2 - \nu^2} - 1 \right) (\Sigma_t - \nu\mu_0) \right] \exp(-\nu x), \qquad (3.29)$$

where ν is the root of the equation

$$(\Sigma_S/2\nu) \ln \frac{\Sigma_t + \nu}{\Sigma_t - \nu} = 1, \qquad (3.30)$$

Σ_t is the total macroscopic interaction cross section, Σ_S is the macroscopic scattering cross section of the medium, and x is the distance from the source. The analytic expression for the case of a cosinusoidal plane source is

$$\varphi^{\,ac}(x) = \int_0^1 \mu_0 \varphi^{\,ac}(x; \mu_0) d\mu_0$$

$$= \left[\Sigma_t/\nu \ln \left(\frac{\Sigma_t}{\Sigma_t - \nu} - 1 \right) \middle/ \left(\frac{\Sigma_S \Sigma_t}{\Sigma_t^2 - \nu^2} - 1 \right) \right] \exp(-\nu x). \qquad (3.31)$$

In this case we can find analytic expressions for the relative sensitivity of $\varphi^{\,ac}$ to Σ_t and Σ_S. For a cosinusoidal plane source, for example, we have

$$p_{\varphi\Sigma_t}^{\,ac} = \frac{\partial\varphi/\varphi}{\partial\Sigma_t/\Sigma_t} = -\frac{\nu\Sigma_t\Sigma_S}{\nu^2 - \Sigma\Sigma_a} x - \beta, \qquad (3.32)$$

$$p_{\varphi\Sigma_S}^{\,ac} = \frac{\partial\varphi/\varphi}{\partial\Sigma_S/\Sigma_S} = t\nu x + \delta, \qquad (3.33)$$

where

$$\beta = \frac{2\Sigma_t\Sigma_S\nu^2}{\nu^2 - \Sigma_t\Sigma_a} - \frac{m\Sigma_t\Sigma_S}{\nu^2 - \Sigma_t\Sigma_a} - 1, \qquad (3.34)$$

$$\delta = t - m + \frac{2\nu^2 - \Sigma_t\Sigma_S}{\nu^2 - \Sigma_t\Sigma_a}, \qquad (3.35)$$

$$t = \frac{\Sigma_t^2 - \nu^2}{\Sigma_t\Sigma_S},$$

$$m = \frac{\nu(\Sigma_t + \nu)}{\Sigma_t\Sigma_S(\Sigma_t/\nu) \ln[\Sigma_t/(\Sigma_t - \nu)] - 1},$$

$$\Sigma_a = \Sigma_t - \Sigma_S. \qquad (3.36)$$

Figure 36. Comparison of experimental data on the fast-neutron distribution in an iron–water shield with results calculated by the ANISN and NRN program (the hatched layers are iron).

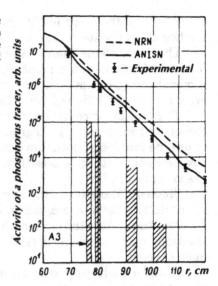

We see that the relative sensitivity of the flux density to a variation in the cross sections increases linearly with the shielding thickness. The relative sensitivity to the total cross section is negative; in other words, an increase in the value of Σ_t leads to a decrease in the flux density φ_0^{as}. A variation of the scattering cross section Σ_S acts in the opposite direction:

$$\frac{\partial \varphi_0 / \varphi_0}{\partial \Sigma_S / \Sigma_S} > 0 .$$

Errors in semiempirical methods

The basic semiempirical method for shielding calculations is the removal-diffusion method, which describes the propagation of fast neutrons by means of removal cross sections, while the propagation of intermediate-energy and thermal neutrons is described through a solution of a diffusion equation. As an example, Fig. 36 compares[153] experimental data on the distribution of fast neutrons with $E \gtrsim 2.5$ MeV in an iron–water mockup of the shield for a pressurized water reactor with the corresponding calculated results, found for a one-dimensional spherical geometry by the S_N method in the ANISN program[130] and by a semiempirical removal-diffusion method by a multigroup NRN program.[154]

We see that the deviation of the results found by the semiempirical method from the experimental data and the results calculated by the S_N method (which are in close agreement) increases with distance from the core; the discrepancy amounts to a factor of 2.5 in this problem at a distance ~50 cm from the core. The data calculated by the NRN program are higher than the experimental data; this is not a typical result for semiempirical calculations, which tend to underestimate the result.

For the full-scale shields of nuclear reactors (with an attenuation factor of $10^{10}-10^{12}$), the error in the shielding calculations by the semiempirical method may be quite substantial.

This error stems from (1) the error in the numerical integration of the influence function of the point source over the volume, (2) the error in the removal cross sections, (3) the error in the other neutron constants, and (4) the error in the calculation of the diffusion component of the neutron flux.

The first error component depends strongly on the particular scheme chosen for the numerical integration. One of the most economical schemes is to use the method of optimum quadrature (cubature) coefficients (the Korobov method).[155] Its error is estimated to be $(\ln K/K)^2$, where K is the number of spatial nodes.

Studies of the error in the numerical integration over the source volume by the KUB-2, PROPARA, etc., programs are described in Ref. 156. It was shown in these studies that the error of the numerical integration can increase substantially with increasing optical dimensions of the source, e.g., the core of a nuclear reactor.

To find a qualitative estimate of the sensitivity of the results of shielding calculations to error in the removal cross sections, we consider the simplest case, of the distribution of removal neutrons of rather high energy in a homogeneous medium from a plane source. In this case the distribution of the neutron flux density is written in terms of the removal cross sections Σ_{rem}:

$$\varphi(x) = B \exp(-\Sigma_{rem} x), \qquad (3.37)$$

where B is a constant.

The relative sensitivity of $\varphi(x)$ to Σ_{rem} is

$$P_{\varphi\Sigma_{rem}} = \frac{\partial\varphi/\varphi}{\partial\Sigma_{rem}/\Sigma_{rem}} = -\Sigma_{rem} x. \qquad (3.38)$$

It follows that the relative sensitivity increases in absolute value with distance from the source. Furthermore, the error of the calculated results is also proportional to the error in the removal cross sections; we know that this error can be quite large, ranging up to 30%.

The error of a calculation of the diffusion component of the neutron flux density depends strongly on the ratio of the fluxes of fast neutrons and intermediate-energy neutrons in the source spectrum and also on the media considered in the problem. Although there is a significant error in the diffusion approximation in the solution of shielding calculation problems (see Sec. 3.1; the method of spherical harmonics), the error in the diffusion component of the neutron flux is determined essentially entirely by the error in the removal cross sections used for the shielding in thermal-neutron reactors, especially if the shielding contains hydrogenous materials, in which case fast neutrons are the leading component.

For the shielding for a fast-neutron reactor, where the role of removal neutrons is unimportant, the error of the calculation method is determined primarily by the error of the diffusion approximation and, in practical calculations, even by the error of the diffusion-return approximation. The error of the latter approximation depends on the number of groups in the neutron energy distribution over which the cross sections are averaged and on other calculation conditions.

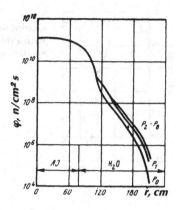

Figure 37. Comparison of calculations in the $S_{16}P_L$ approximation for various L for the fast-neutron distribution in the water shield of a water-cooled, water-moderated reactor.

It should be noted that an iterative approach (for a repeated averaging of the constants) should be taken to the solution of the multigroup neutron transport equation. This comment applies, in particular, to cases in which the diffusion-return approximation is used, by virtue of the uncertainty in the intragroup spectrum. In practical calculations, frequently only the first iteration is used, with the constants being found under the assumption that a Fermi spectrum applies [$\varphi(u) = $ const, where u is the lethargy]. When this approach is taken, a small number of groups leads to an overestimate of the results, while a large number of groups leads to a significant underestimate. The optimum group structure—which minimizes the error of the diffusion-return calculations—would have about 25 groups, corresponding to a width $\Delta u \approx 0.6$ for the group interval.

Errors in numerical methods for solving the transport equation

As a rule, studies of real shields are based on numerical methods for solving the transport equation. The total set of input parameters can be quite large. In such cases, studies can be carried out primarily for simple shields which would reflect the basic features of the real shields of nuclear-engineering installations. The simplicity of a shield eliminates complex effects of a superposition (influence) of certain parameters on others and makes it possible to distinguish primary and secondary parameters. Foremost among the simple shields are homogeneous media and heterogeneous shields with a simple layout.

Let us first examine the possible errors which would arise in numerical methods because of the approximate description of the scattering characteristic.

To illustrate the scale of these errors we show in Fig. 37 results calculated in the $S_{16}P_L$ approximation ($L = 0,2,8$) on the propagation of neutrons with an energy $E > 10$ MeV in water.[153] Knowledge of the distribution of these neutrons is important, in particular, in the calculation of the oxygen activity of a heat exchanger. The approximations from P_2 to P_8 yield similar results.

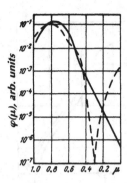

Figure 38. Dashed line, comparison of results calculated on the angular distribution of the flux density of fast neutrons behind a hydrogen plate with a thickness of 1 mean free path in the P_8 approximation of the scattering characteristic; solid line, exact data.

When we go to the P_1 approximation of the scattering characteristic, we find that the results at large distances become several times smaller; when we go to the P_0 approximation, we find that the results are reduced significantly over the entire volume of the shield.

From data of this type we can draw conclusions about the extent to which the calculated results are sensitive to errors in the description of the scattering characteristic. In the case just discussed, for example, the results of the calculations are relatively insensitive to errors in the P_2 through P_8 moments of the scattering characteristic, while they are quite sensitive to error in the P_0 and P_1 moments.

The error in shielding calculations which stems from the approximate description of the neutron scattering characteristic increases with increasing neutron energy E, because of an increase in the scattering anisotropy with an increase in E.

The error due to the approximate description of the scattering characteristic is largest in calculations on neutron propagation in hydrogenous materials. The angular dependence of the differential cross sections for neutron scattering from group to group for such materials is a complex function with peaks, which can be described only poorly by a finite series of Legendre polynomials (Fig. 17; Ref. 157). Even a higher-order approximation of the scattering characteristic yields a poor description of the exact picture in such cases. This circumstance cannot fail to affect calculations on neutron propagation in hydrogenous media. Figure 38 shows results of calculations by the S_N method of the angular flux density of fast neutrons behind a hydrogen plate with a thickness of 1 mean free path, bombarded by a beam of neutrons with energies of 13.5–15.9 MeV. The results of these calculations in the P_8 approximation of the scattering characteristic are compared with data from exact calculations, based on a discrete angular representation of the scattering characteristic.[158] The calculations in the P_8 approximation lead to oscillatory, and frequently negative, values of the angular flux density of the neutrons. For a 10-cm thickness of water, the agreement of the data is slightly better than for hydrogen; this comment also applies to reflected neutrons (Fig. 18). With decreasing neutron energy, the accuracy of the P_8 calculations becomes better.

The switch to heterogeneous shields may result in some cancellation of the approximate nature of the description of the actual angular dependence

Figure 39. Distributions of fast neutrons with energies $E > 2.5$ MeV in the iron-water shield of a water-cooled, water-moderated reactor. Dashed line, calculated in a one-dimensional geometry by the S_N method in the $S_{16}P_3$ approximation; solid line, in the $S_{16}P_1$ approximation (the ANISN program) (Ref. 153).

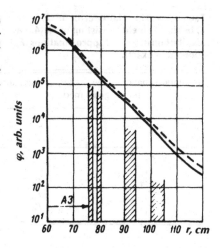

Table XXIII. Difference (%) in the results calculated for the flux density of fast neutrons at a distance of 80 cm from the active zone.

E,MeV	Approximation of the scattering characteristic, PL				
	P_1	P_2	P_3	P_5	P_7
$\geqslant 2.5$	− 47.9	− 6.36	− 0.21	− 0.05	—
$\geqslant 1.4$	− 48.1	− 6.22	− 0.13	− 0.03	—
0.1–1.4	− 46.2	− 5.92	− 0.15	0.01	—

of the characteristic curve for neutron scattering by different materials.

As a result, the sensitivity of the results calculated on neutron fields in heterogeneous shields to an error in the description of the scattering characteristic usually turns out to be weaker than for homogeneous shields (for given attenuation levels; Fig. 39). In the example shown, the switch from the P_3 approximation of the scattering characteristic to the P_1 approximation results in a decrease in the neutron flux density by a factor of about 1.7 at a distance of about 50 cm from the core.

Table XXIII shows the deviations of the results calculated by the discrete-ordinates method on the flux density of fast neutrons ($E \geqslant 2.5$ MeV) behind an iron-water shield (10 cm Fe + 30 cm H_2O + 10 cm Fe + 30 cm H_2O) of a water-cooled, water-moderated reactor, found in the $2D_7P_7$ approximations, from the results in the $2D_{12}$ approximation taken to be exact for the flux density, with a discrete representation of the scattering characteristic.[144]

Analysis of these studies shows that, for attenuation levels of 10^5–10^6 for the flux density of fast neutrons in the iron-water shield, the P_2 and P_3 approximations of the scattering characteristic are sufficient in practice for calculations by the discrete-ordinates method.

In calculation problems where the role of high-energy neutrons is inconsequential, e.g., in calculations on the shielding of fast reactors, the sensitivity of the calculated results to errors in the description of the scattering char-

Table XXIV. The relative error $\delta P/P$, %, in calculations of the tissue absorbed dose absorption rate P, behind a shield consisting of Na(4.6 m) + Fe(0.9 m) for a neutron source with the FTR energy distribution in several approximations (P_L) of the scattering characteristic (the P_3 approximation is adopted as exact).

P_L: approximation of the scattering characteristic for the given medium		Calculations of $\delta P/P$, %, by the SWANLAKE and ANISN programs	
Na	Fe	Estimates from linear perturbation theory	Calculations by the direct-substitution method
P_2	P_2	—	− 0.03
P_1	P_1	− 18	− 13.8
P_2	P_1	—	− 16
P_0	P_0	− 278	− 88
P_1	P_0	—	− 27

Table XXV. Coefficient of the nonuniformity of the angular flux density of neutrons behind an iron–water shield with a thickness of 80 mm in the $2D_7P_L$ approximation.

E,MeV	P_1	P_2	P_3	P_5	P_7	Exact
⩾4	3.42	4.29	4.41	4.53	4.54	4.58
⩾2.5	3.07	3.49	3.93	4.01	4.02	4.28
⩾1.4	2.79	3.23	3.57	3.62	3.62	3.75

acteristic is weak. Analysis of the data on problem 7 in Table V, shown in Table XXIV, for a large amount of sodium (4.6 m) surrounded by a layer of iron (0.9 m), bombarded by neutrons from the FTR, the P_1 approximation of the scattering characteristic is quite sufficient for calculations of the absorbed dose rate. The sensitivity to the P_2 moment of the scattering characteristic is quite weak.

When we turn to the angular distribution of the neutron flux density we find a slightly different situation. In such cases, the requirements imposed on the scattering characteristic for the calculation of the correct results become more stringent. Table XXV shows data on the nonuniformity coefficient of the angular flux density of neutrons [the ratio $\varphi(\mu = 1)$ in the forward direction to the average value of the angular flux, averaged over all angles]; this coefficient is a measure of the degree of anisotropy.[144]

We see that a correct description of the angular dependence of the neutron flux density at an attenuation $\sim 10^6$ requires at least the P_3 approximation of the scattering characteristic. At a greater attenuation level, a higher-order approximation of the scattering characteristic becomes necessary.

The sensitivity of the results calculated on the angular flux density of the neutrons is thus considerably greater than in calculations on the neutron flux density.

By themselves, large errors in the calculated results and a pronounced sensitivity of these results to the harmonics of the scattering cross section, without a comparison with corresponding data for other partial cross sec-

Table XXVI. Components of the relative error $\delta\bar{R}_k/R_k$ (%) related to the values of the cross sections and relative sensitivities R to the interaction cross sections.

Parameter X_n	Relative sensitivity R_k to X_n				Components of the relative error $\delta\bar{R}_k/R_k$ (%) and indices of the energy groups which determine the error (in the denominator)	
	$f+1/E$		INC		$f+1/E$	INC
Cross section	i[a]	u[a]	i	u	i	u
			Sodium (4.6 m). The functional φ_F			
$\sigma_{c+el+in}$	$-1.72+1$[b]	$-1.71+1$	$-1.73+1$	$-1.74+1$	191	194
σ_c	$-1.27-1$	$-1.52-1$	$-1.79-1$	$-2.21-1$	8/1	11/1
σ_{in}	-8.51	-8.37	-8.35	-8.35	$166/1-4$	$166/1-4$
σ_{el}	-8.49	-8.63	-8.85	-8.85	$95/1-4$	$100/1-4$
σ_{el0}	$-1.38+1$	$-1.40+1$	$-1.40+1$	$-1.46+1$		
σ_{el1}	4.72	4.57	4.56	4.50		
σ_{el2}	$5.36-1$	$7.39-1$	$5.43-1$	$7.88-1$		
σ_{el3}	$3.61-2$	$1.22-1$	$4.09-2$	$1.48-1$		
			The functional P_{equ}			
$\sigma_{c+el+in}$	$-1.20+1$	$-1.20+1$	$-1.25+1$	$-1.25+1$	49	56
σ_c	$-8.08-1$	$-7.86-1$	$-9.01-1$	$-8.90-1$	$5/>14$	$6/>14$
σ_{in}	$-4.00-1$	$-4.70-1$	$-1.34-2$	$-1.46-2$	$9/2-4$	0
σ_{el}	$-1.08+1$	$-1.07+1$	$1.16+1$	$-1.17+1$	$48/>1$	$56/>6$
σ_{el0}	$-1.13+1$	$-1.13+1$	$-1.19+1$	$-1.19+1$		
σ_{el1}	$5.07-1$	$5.51-1$	$2.75-1$	$2.77-1$		
σ_{el2}	$2.78-3$	$2.40-2$	$1.68-2$	$1.67-3$		
σ_{el3}	$2.52-4$	$1.16-3$	$2.30-3$	$-4.40-5$		
			The functional P			
$\sigma_{c+el+in}$	$-1.20+1$	$-1.20+1$	$1.25+1$	$1.26+1$	49	57
σ_c	$-7.84-1$	$-7.86-1$	$-8.70-1$	$-8.59-1$	$5/>14$	$6/>14$
σ_{in}	$-3.79-1$	$-4.70-1$	$-1.35-2$	$-1.46-2$	$9/2-4$	0
σ_{el}	$-1.08+1$	$-1.07+1$	$-1.16+1$	$-1.17+1$	$48/>1$	$57/>6$
σ_{el0}	$-1.13+1$	$-1.13+1$	$-1.19+1$	$-1.20+1$		
σ_{el1}	$4.97-1$	$5.52-1$	$2.76-1$	$2.79-1$		
σ_{el2}	$1.81-3$	$2.40-2$	$-1.70-4$	$1.68-3$		
σ_{el3}	$2.11-4$	$1.16-3$	$2.30-5$	$-4.42-5$		
			The functional $\varphi_{1/v}$			
$\sigma_{c+el+in}$	$-1.10+1$	$-1.10+1$	$-1.14+1$	$-1.15+1$	49	57
σ_c	-1.77	-1.75	-1.83	-1.82	$5/>20$	$5/>20$
σ_{in}	$-2.37-1$	$-2.70-1$	$-9.22-3$	$-1.01-2$	$5/>2$	$5/>2$
σ_{el}	-9.01	-8.97	-9.60	-9.65	$30/>4$	$32/>4$
σ_{el0}	-9.37	-9.37	-9.82	-9.86		
σ_{el1}	$-3.66-1$	$-3.91-1$	$2.13-1$	$2.15-1$		
σ_{el2}	$-5.74-4$	$1.27-2$	$-2.46-4$	$1.25-3$		
σ_{el3}	$2.02-4$	$-7.13-5$	$2.17-5$	$-4.94-5$		
			Iron (0.5 m). The functional φ_F			
$\sigma_{c+el+in}$	-7.87	-7.93	-7.84	-7.91	48	48
σ_c	$-1.88-2$	$-1.80-2$	$-1.75-2$	$-1.62-2$	0	0

Table XXVI. Cont'd.

σ_{in}	-5.06	-4.77	-4.94	-4.61	$45/1-4$	$45/1-4$
σ_{el}	-2.79	-3.14	-2.88	-3.29	$18/1-4$	$18/1-4$
σ_{el0}	-4.25	-4.67	-4.13	-4.58		
σ_{el1}	1.28	1.16	-1.07	$9.48-1$		
σ_{el2}	$1.84-1$	$3.17-1$	$1.77-1$	$2.97-1$		
σ_{el3}	$-1.43-3$	$4.30-2$	$-1.74-3$	$4.01-2$		

The functional P_{equ}

$\sigma_{c+el+in}$	-1.77	-1.83	-1.40	-1.39	4.4	3.1
σ_c	$-5.64-2$	$-5.15-2$	$-6.28-2$	$-5.35-2$	0	0
σ_{in}	$-1.46-1$	$-1.76-1$	$-1.27-3$	$-1.60-3$	$1.4/2-5$	0
σ_{el}	-1.56	-1.59	-1.34	-1.33	$4.2/4-10$	$3.1/4-10$
σ_{el0}	-1.74	-1.79	-1.41	-1.42		
σ_{el1}	$-1.78-1$	$-1.92-1$	$7.69-2$	$8.44-2$		
σ_{el2}	$-4.04-3$	$7.95-3$	$-9.70-4$	$9.60-4$		
σ_{el3}	$-6.24-4$	$-1.47-3$	$2.10-5$	$-4.80-5$		

The functional P

$\sigma_{c+el+in}$	-1.50	-1.53	-1.20	-1.15	3.5	2.6
σ_c	$-6.56-2$	$-5.96-2$	$-7.17-2$	$-6.07-2$	0	0
σ_{in}	$-1.08-1$	$-1.34-1$	$-7.64-4$	$-9.75-4$	$1.1/2-7$	0
σ_{el}	-1.32	-1.38	-1.13	-1.09	$3.3/4-10$	$2.6/4-10$
σ_{el0}	-1.46	-1.53	-1.19	-1.16		
σ_{el1}	$1.43-1$	$1.57-1$	$5.84-2$	$6.48-2$		
σ_{el2}	$-3.43-3$	$5.70-3$	$-7.17-4$	$5.02-4$		
σ_{el3}	$5.48-4$	$-1.35-3$	$1.61-5$	$-4.10-5$		

The functional $\varphi_{1/v}$

$\sigma_{c+el+in}$	-1.22	-1.17	-2.31	$-7.61-1$	2.6	1.2
σ_c	$-6.41-1$	$-6.20-1$	$-9.34-1$	$-5.67-1$	$1.3/>14$	$1.0/>10$
σ_{in}	$-4.01-3$	$-6.96-3$	$1.50-2$	$-3.55-5$	0	0
σ_{el}	$-5.70-1$	$-5.42-1$	-1.37	$-1.94-1$	$2.2/>7$	$0.7/>9$
σ_{el0}	$-6.05-1$	$-5.82-1$	-1.37	$-2.12-1$		
σ_{el1}	$3.567-2$	$4.07-2$	$-2.63-3$	$1.88-2$		
σ_{el2}	$-1.23-3$	$-2.37-4$	$-1.72-4$	$-2.09-4$		
σ_{el3}	$2.48-4$	$-7.58-4$	$4.02-6$	$6.14-6$		

[a] Isotropic (i) or unidirectional (u) angular distribution of the radiation from the source.
[b] Read a-8 as $a \times 10^{-8}$.

tions, give a poor characterization of the accuracy of calculations. Consequently, in several cases these studies have been carried out in association with studies for various partial cross sections. The only effective way to carry out such studies is to use perturbation theory.

For an illustration of studies of this type, shielding materials characteristic of fast-neutron reactors with resonances in the interaction cross section were selected. This resonance structure is important in problems involving a deep penetration. In the present section we discuss the results of studies only by the authors of the present book for benchmark calculation problems in the examples of a homogeneous shield (problem 1 in Table V; Ref. 110) and a heterogeneous shield (problem 6 in Table V; Ref. 115) of Fe and Na. These calculations were carried out by the ZAKAT + ROZ-11 + ARAMAKO-2F software, with the help of the CORE program for evaluating the cross section component of the error. Tables XXVI and XXVII and Fig. 40 show results of a study of the

Table XXVII. Components of the relative error $\delta\bar{R}_k/R_k$ (%) related to the values of the cross sections and relative sensitivities R to the interaction cross sections.

Parameter, X_n	Relative sensitivity R_k and X_n				Components of relative error $\delta\bar{R}_k/R_k$, % (numerator) and indices of the energy groups which determine the error (denominator)			
	f + 1/E		INC		f + 1/E		INC	
Cross section	Na	Fe	Na	Fe	Na	Fe	Na	Fe
Na(4.6 m) + Fe(0.5 m). The functional φ_F								
$\sigma_{c+el+in}$	−1.72 + 1	−7.98	−1.77 + 1	−8.06	48	183	185	39
σ_c	−1.70 − 1	−2.31 − 2	−2.43 − 1	−2.40 − 2	0.3/4	8.4/1 − 2	12.3/1 − 2	0.3/4
σ_{in}	−8.44	−5.09	−8.51	−5.16	45.5/4	166/1 − 4	168/1 − 4	36.8/4
σ_{el}	−1.71 + 1	−7.96	−1.74 + 1	−8.03	12.2/4	77.4/1 − 4	75.8/1 − 4	12.2/4
σ_{el0}	−2.27 + 1	−9.60	−2.32 + 1	−9.71				
σ_{el1}	4.61	1.35	4.62	1.38				
σ_{el2}	8.29 − 1	2.72 − 1	9.15 − 1	2.84 − 1				
σ_{el3}	1.45 − 1	1.67 − 2	1.78 − 1	1.86				
The functional P_{equ}								
$\sigma_{c+el+in}$	−1.70 + 1	−1.36	−2.02 + 1	−1.22	3	96	112	3.3
σ_c	−5.75 − 2	−7.38 − 2	−2.50 − 1	−8.35 − 2	0.2/10	0.3/1 − 2	0	0.2/10
σ_{in}	−4.00	−1.76 − 2	−2.23 − 1	−6.84 − 5	0.3/4 − 5	78.7/2 − 4	4.0/6	0
σ_{el}	−1.69 + 1	−1.28	−2.02 + 1	−1.13	2.9/6 − 10	54.5/1 − 10	112/6 − 10	3/18 − 10
σ_{el0}	−1.97 + 1	−1.34	−2.09 + 1	−1.16				
σ_{el1}	2.45	6.05 − 2	7.53 − 1	3.15 − 2				
σ_{el2}	2.68 − 1	5.14 − 5	1.71 − 2	9.92 − 5				
σ_{el3}	3.74 − 2	6.95 − 6	3.98 − 4	7.40 − 7				
The functional P								
$\sigma_{c+el+in}$	−1.69 + 1	−1.21	−2.02 + 1	−1.21	3	96	112	3.3
σ_c	−5.44 − 2	−8.63 − 2	−2.57 − 2	−1.01 − 1	0.2/10	0.3/1 − 2	0	0.2/10
σ_{in}	−3.76	−9.64 − 3	−2.16 − 1	−3.46 − 5	0.3/4 − 5	78.7/2 − 4	4.0/6	0
σ_{el}	−1.69 + 1	−1.12	−2.02	−1.71	29/6 − 10	54.5/1 − 10	112/6 − 10	3/8 − 10
σ_{el0}	−1.95 + 1	−1.16	−2.09 + 1	−1.14				
σ_{el1}	2.34	−1.38 − 2	7.46 − 1	−2.70 − 2				
σ_{el2}	2.47 − 1	−5.37 − 5	1.66 − 2	1.05 − 4				
σ_{el3}	3.39 − 2	4.32 − 6	3.63 − 4	6.71 − 7				
The functional $\varphi_{1/v}$								
$\sigma_{c+el+in}$	−1.67 + 1	−2.51	−1.87 + 1	−6.25	4	85	99	10
σ_c	−5.80 − 2	−8.82 − 1	−8.79 − 2	−1.57	2.6/ > 10		0	4.6/ > 18
σ_{in}	−3.14	−4.63 − 4	−1.46 − 1	8.47 − 7	0	62.8/1 − 6	3.0/6	0
σ_{el}	−1.66 + 1	−1.63	−1.86 + 1	−4.68	3.3/10	56.6/1 − 11	99.0/6	8.7/ > 10
σ_{el0}	−1.89 + 1	−1.67	−1.92 + 1	−4.73				
σ_{el1}	2.03	−2.27 − 2	6.27 − 1	5.29 − 2				
σ_{el2}	1.97 − 1	−1.13 − 4	1.16 − 2	−3.53 − 5				
σ_{el3}	2.56 − 2	3.8 − 7	1.98 − 4	2.04 − 7				

Na(4.6 m) + Fe(0.9 m). The functional φ_F

				φ_F				
$\sigma_{c+el+in}$	-1.72 + 1	-1.45 + 1	-2.04 + 1	-1.65 + 1	181	87	156	86
σ_c	-1.72 - 2	-3.79 - 2	0	0	1/2	0	0	0
σ_{in}	-8.43	-8.99	-5.62	-8.49	164/1 - 4	83/3 - 4	108/1 - 4	80/4
σ_{el}	-1.70 + 1	-1.45 + 1	-2.04 + 1	-1.65 + 1	78/1 - 4	25/4	112/1 - 4	32/4
σ_{el0}	-2.27 + 1	-1.73 + 1	-2.18 + 1	-1.82 + 1				
σ_{el1}	4.62	2.20	1.42	1.72	—	—	—	—
σ_{el2}	8.38 - 1	5.52 - 1	0	0	—	—	—	—
σ_{el3}	4.48 - 1	4.82 - 2	0	0				

The functional P_{equ}

$\sigma_{c+el+in}$	-1.93	-1.69 + 1	-2.02 + 1	-1.76	93	5.7	116	6.6
σ_c	-1.31 - 1	-5.38 - 2	-2.47 - 1	-1.28 - 1	-1/1 - 2	0.6/10	0.5/9 - 10	0.5/10
σ_{in}	-8.40 - 3	-5.72	-2.15 - 1	-2.95 - 5	74/2 - 4	0	4.2/6	0
σ_{el}	-1.80	-1.69 + 1	-2.02 + 1	-1.63	56/2 - 10	5.7/10	116/6 - 4	6.6/18
σ_{el0}	-1.86	-1.95 + 1	-2.09 + 1	1.69				
σ_{el1}	6.27 - 3	2.32	7.46 - 1	4.09 - 2	—	—	—	—
σ_{el2}	1.00 - 4	3.43 - 1	1.66 - 2	-8.62 - 5	—	—	—	—
σ_{el3}	4.33 - 2	3.33 - 2	3.59 - 4	1.02 - 7	—	—	—	—

The functional P

$\sigma_{c+el+in}$	-1.69 + 1	-1.81	-2.02 + 1	-1.72	93	5.7	116	6.6
σ_c	-5.27 - 2	-1.40 - 1	-2.48 - 2	-1.36 - 1	1/1 - 2	0.6/10	0.5/9 - 10	0.5/10
σ_{in}	-1.69 + 1	-1.67	-2.13 + 1	-1.72 - 5	74/2 - 4	0	1.2/6	0
σ_{el}	-3.63	-5.04 - 3	-2.02 + 1	-1.59	56/2 - 10	5.7/10	116/6 - 4	6.6/10
σ_{el0}	-1.94 + 1	-1.72	-2.09 + 1	-1.63				
σ_{el1}	2.27	5.20 - 2	7.45 - 1	3.84 - 2	—	—	—	—
σ_{el2}	2.35 - 1	-1.96 - 5	1.64 - 2	-9.53 - 5	—	—	—	—
σ_{el3}	3.19 - 2	2.49 - 6	3.48 - 4	6.35 - 8	—	—	—	—

The functional $\varphi_{1/v}$

$\sigma_{c+el+in}$	-1.69 + 1	-1.66	-2.02 + 1	-1.71	90	3.9	116	8.7
σ_c	-5.10 - 2	-6.74 - 1	-2.49 - 2	-6.73 - 1	1/1	3.0/ >10	0.7/8 - 10	0.5/10
σ_{in}	-3.51	-1.09 - 3	-2.10 - 1	-3.65 - 6	70/2 - 4	0	4.1/6	0
σ_{el}	1.09 + 1	-9.88 - 1	-2.02 + 1	-1.04	56/2 - 10	2.5/10	116/6 - 10	2.9/10
σ_{el0}	-1.94 + 1	-1.02	-2.10 + 1	1.07				
σ_{el1}	2.22	2.98 - 2	7.44 - 1	2.65 - 2	—	—	—	—
σ_{el2}	2.23 - 1	-1.44 - 4	1.62 - 2	-1.03 - 4	—	—	—	—
σ_{el3}	3.01 - 2	4.92 - 7	3.34 - 4	-3.01 - 8	—	—	—	—

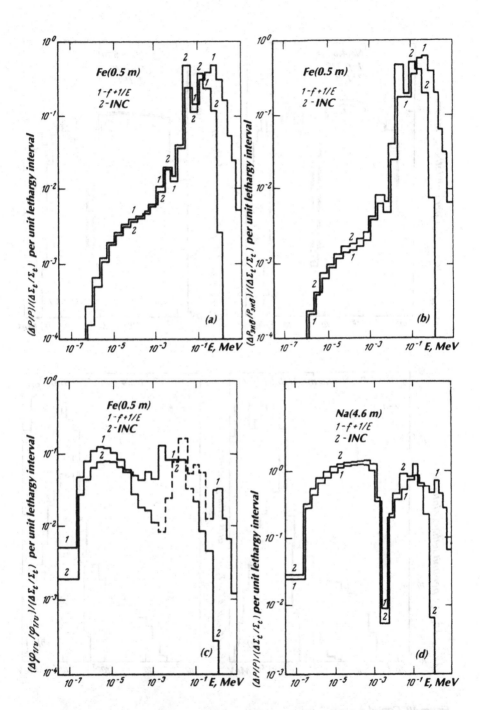

Figure 40. Energy dependence of the relative sensitivity of the equivalent dose rate
P_{equ} [(b), (e), (h), (k), (n), (q)], of the absorbed dose rate P in tissue [(a), (d), (g), (j), (m),
and (p)] and readings of a $1/v$ detector, $\varphi_{1/v}$ [(c), (f), (i), (l), (o), (r)] behind
homogeneous and heterogeneous shields of iron and sodium to the total cross
section for unidirectional plane neutron sources with various energy distributions.
Solid lines, negative values; dashed lines, positive values. For the heterogeneous
media, 1 and 2 specify the sensitivity of the result to Σ_t for Na and Fe, respectively.

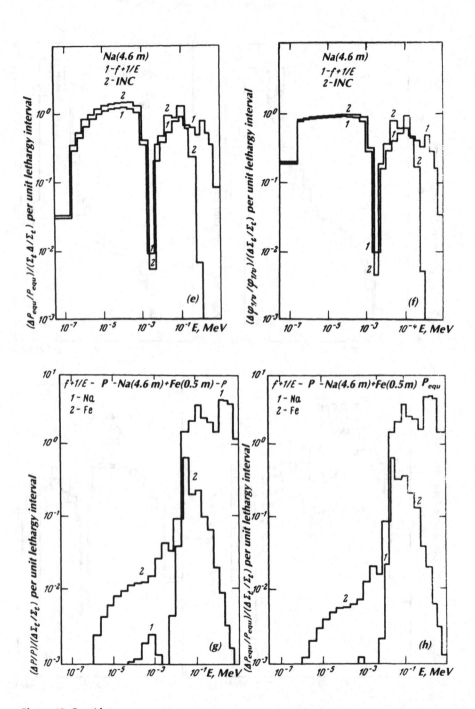

Figure 40. Cont'd.

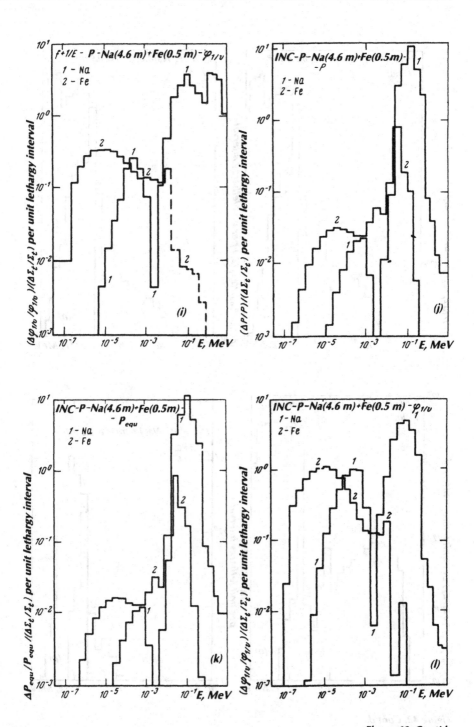

Figure 40. Cont'd.

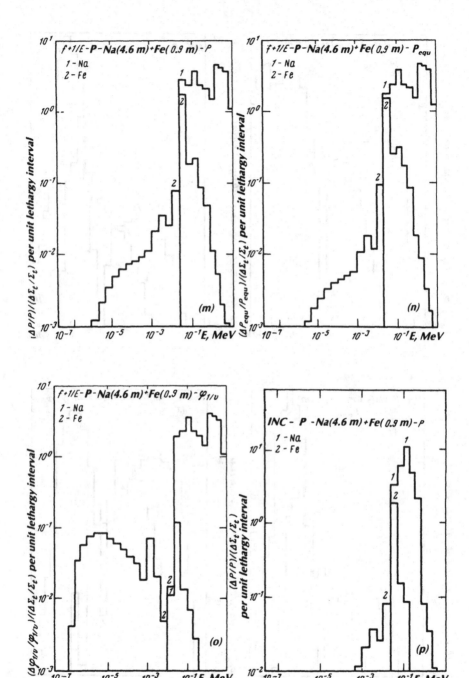

Figure 40. Cont'd.

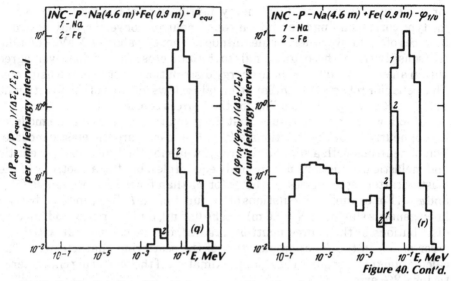

Figure 40. Cont'd.

sensitivity to the total and partial cross sections for the neutron-matter inter-action of calculations of the flux density (φ_F) of fast neutrons with energies $E \geqslant 1.4$ MeV, on the equivalent dose rate P_{equ}, on the absorbed dose rate in tissue, P, and on the readings of a $1/v$ detector, $\varphi_{1/v}$, behind shields. Also shown here are the corresponding errors in these functionals for a plane geometry with plane isotropic and unidirectional sources of neutrons with various energy distributions ($f + 1/E$ and the energy distribution corre-sponding to an intermediate-neutron converter) for shields of Na (4.6 m), Fe (0.5 m), Na (4.6 m) + Fe (0.5 m), and Na (4.6 m) + Fe (0.9 m).

We can draw some conclusions from these results. The angular distribu-tion of the source neutrons has little effect on the relative sensitivity or the energy dependence of the relative sensitivity of these functionals to the total cross section ($\sigma_{c+el+in}$), the absorption cross section (σ_c), the elastic cross section (σ_{el}), or the inelastic cross section (σ_{in}). For example, for homogeneous shields of Fe (0.5 m) and Na (4.6 m) (Table XXVI) the relative sensitivities of φ_F, P_{equ}, P, and $\varphi_{1/v}$ to these partial cross sections are the same, within an error of 20%, for plane sources with isotropic and unidirectional angular distributions. This result is explained on the basis that the spatial depend-ence of the energy flux density of the neutrons becomes independent of the angular distribution of the source neutrons, within a constant factor, at a certain distance from the source, as was mentioned in Sec. 3.2. Physically, this result shows that the picture of neutron transport is similar for sources with different angular distributions and makes it possible to evaluate the constant component of the error for plane sources with only a single angular distribution of the radiation, e.g., a unidirectional distribution.

The energy distribution of the source neutrons, examined for the particu-lar cases of an $f + 1/E$ source and a source corresponding to an intermediate-neutron converter, has a greater effect than the angular distribution on the energy dependence of the relative sensitivity. For example, the energy depen-dences of the relative sensitivity of φ_F, P_{equ}, P, and $\varphi_{1/v}$ behind the shields studied, to both the total cross section (Fig. 40) and the partial cross sections, for plane sources with neutron energy distributions of the $f + 1/E$ type and the type corresponding to an intermediate-neutron converter, may have differences of up to two orders of magnitude at energies above 1 MeV. The relative sensitivities of these functionals to the total cross section, however, agree within an error of 30% for these energy distributions of the source neutrons. This behavior reflects the similar attenuation laws for each of the functionals φ_F, P_{equ}, P, and $\varphi_{1/v}$ in shielding for the sources considered.

It follows from these results that the primary processes for the moderation of neutrons for the functionals P_{equ}, P, and $\varphi_{1/v}$ are the elastic moderation of neutrons with a slight anisotropy, while for the functional φ_F elastic and inelastic interactions play roughly equal roles, but the anisotropy of the elastic-scattering processes is greater for φ_F than for the functionals P, P_{equ}, and $\varphi_{1/v}$. For example, calculations of the functionals P, P_{equ}, and $\varphi_{1/v}$ behind homogeneous shields of Na (4.6 m) and Fe (0.5 m), for the energy and angular distributions of the source neutrons studied, can be carried out within an error of 2% in the P_1 approximation of the scattering characteristic. The calculations of φ_F require the P_3 approximation of the scattering characteristic for a 2% error.

The role of the capture process is more important for the readings of a $1/v$ detector, $\varphi_{1/v}$, than for the other functionals, but it does not become the governing factor in the shaping of the readings of the $1/v$ detector. In all cases, as expected, inelastic scattering play a greater role for a source energy distribution $f + 1/E$ than for a source energy distribution corresponding to an intermediate-neutron converter, since the latter has a softer energy distribution of emitted neutrons.

Information of importance for studies comes from the energy dependence of the relative sensitivity of φ_F, P_{equ}, P, and $\varphi_{1/v}$ behind shielding to the total cross section. These quantities characterize the roles played by various energy ranges in the moderation of the neutrons. These relative sensitivities have some general features which are related to features of the energy dependence of the neutron-matter interaction cross section (Fig. 40).

For example, the dip in the energy dependence of the relative sensitivity of the functionals P_{equ}, P, and $\varphi_{1/v}$ behind a Na shield (4.6 m) to the total cross section at energies in the range 2.15–4.65 keV corresponds to a maximum in the total cross section for the interaction of neutrons with sodium. The small magnitudes of the relative sensitivities for this dip show clearly that the interactions of neutrons with Na described by the interaction cross section in this energy interval are insignificant. For example, an error of $\pm 50\%$ in this cross section for the interaction of neutrons with Na leads to an error of no more than $\pm 0.4\%$, according to expression (1.17), in calculations of these functionals. The approximate agreement of the energy dependence of the relative sensitivity for each of the functionals, P_{equ}, P, and $\varphi_{1/v}$ over the energy interval 10^{-5}–10^{-3} MeV is due to the essentially constant values of the total cross section for the interaction of neutrons with sodium and the approximate agreement of the physical picture for the transport of neutrons with these energies E.

Studies of the energy dependence of the relative sensitivity of P_{equ}, P, and $\varphi_{1/v}$ behind 0.5 m of iron to the total cross section showed, in particular, that the functionals P_{equ} and P are formed primarily by interactions of neutrons with energies in the range 2.15×10^{-2}–1.4 MeV, while $\varphi_{1/v}$ is formed primarily by interactions of neutrons with energies in the ranges 4.65×10^{-7}–10^{-4} and 10^{-3}–10^{-1} MeV. The role played by the interaction of neutrons with Fe is quite different in the two latter energy intervals. While an increase in the cross section in the first energy interval reduces the detector readings $\varphi_{1/v}$, an increase in the cross section of the second interval increases these readings. The reason is that in the former case there is a decrease in the neutron energy flux density $\varphi(E)$ at energies where the detection efficiency is high, while in the latter case $\varphi(E)$ decreases in an energy region where the detection efficiency is low, but $\varphi(E)$ increases in an energy region where the detection efficiency is high.

For the heterogeneous Na and Fe shields studied, the minimum in the total neutron-iron interaction cross section, in the energy group 21.5–46.5 keV, corresponds to a maximum in the energy dependence of the relative sensitivity of P_{equ} and P to the total cross section. The role played by this minimum in the cross section increases with the Fe thickness, at an Fe thickness of about 0.9 m, it essentially completely determines the interactions of neutrons with iron. This result points to the need for a correct account of the resonance structure of the neutron-iron cross section for the energy region 21.5–46.5 keV in constructing a group cross section for the interaction. A consequence of the leading role played by neutrons with energies of 21.5–46.5 keV is the fact that the energy distributions of the relative sensitivity of P_{equ}, P, and $\varphi_{1/v}$ to the total Na cross section at energies above 10^{-2} MeV are essentially independent of the Fe thickness. Consequently, the Fe in these heterogeneous shields serves as a filter which cuts out low-energy neutrons.

Let us examine the structure and sources of the errors in the determination of the functionals φ_F, P_{equ}, P, and $\varphi_{1/v}$ behind these shields which stem from uncertainties in the total and partial neutron cross sections used in the calculations (the cross sections for radiative capture, elastic scattering, and inelastic scattering).

The cross section component of the errors in these functionals can be subdivided further into the following components: (1) the component of the error which is due to uncertainties in the microscopic cross sections of the individual isotopes from which the group cross sections are constructed (this component, which is independent of the calculated composition, is determined by the accuracy of the initial information for the calculation of the macroscopic cross sections of the given medium); (2) the component of the error which is due to the macroscopic components and which stems from the error in the averaging of the cross sections over broad energy groups with a flux weight in the standard form (a fission spectrum f, a $1/E$ spectrum, or a Maxwellian spectrum).

The component of the error due to the approximate account of the elastic moderation of neutrons in the transition from the given group to the next group deserves special mention.

In Chap. 2, Sec. 1 we discussed the question of obtaining neutron data from which the files of evaluated data are constructed from libraries of group microscopic cross sections. In the same place we discussed estimates of the errors of the estimated data and of the group macroscopic cross sections for the individual isotopes.

The LUND library of errors in the group microscopic cross sections which have been developed, and which is used in the INDÉKS system (Sec. 2.2), can be used to estimate the component of the error in a shielding calculation which stems from the first of these sources of error: the microscopic data.

Tables XXVI and XXVII show estimates of the errors in the calculated predictions of the functionals φ_F, P_{equ}, and $\varphi_{1/v}$ for various versions of the shield. The components of the errors which stem from the various partial interactions of neutrons with nuclei of the shielding material are estimated from (1.18).

For the homogeneous shields the uncertainty in the capture cross sections is insignificant according to estimates of the errors in the functionals φ_F, P_{equ}, and $\varphi_{1/v}$. The error in the φ_F calculation is determined primarily by uncertainties in the inelastic-scattering cross sections, which essentially completely determine the loss of neutrons from the high-energy region.

For homogeneous shields the individual components of the error (c,el,in) and, correspondingly, their sum (c + el + in) for the functionals considered depend only weakly on the energy distribution of the source neutrons, differing from each other by a factor of no more than 1.5. The highly accurate prediction of the functionals for the iron shield, in contrast with the sodium shield, is attributed to a difference in the factors by which the radiation is attenuated by the shield for these functionals in this example (for sodium, these values are several orders of magnitude larger) and to the smaller errors in the cross section for the interaction of neutrons with Fe.

For heterogeneous shields (Table XXVII), as in the case of homogeneous media, the primary sources of errors in the functionals φ_F, P_{equ}, and $\varphi_{1/v}$ are the uncertainties in the cross sections for elastic and inelastic scattering.

For φ_F the components of the errors which are due to the uncertainties in the individual interaction processes (c,el,in) and the total error (c + el + in) are essentially equal to the corresponding quantities for the corresponding homogeneous shields. The explanation here is that φ_F is determined by neutrons with energies in the range 1.4–10.5 MeV, for which the attenuation is described well by an exponential thickness dependence.

For P_{equ} and the readings of the $1/v$ detector, $\varphi_{1/v}$, and for a neutron source with an $f + 1/E$ energy distribution, the errors due to the uncertainties in the cross sections for inelastic scattering in sodium are roughly ten times as large as for a homogeneous shield.

In the case of a source corresponding to an intermediate-neutron converter, the total error for the functionals P_{equ} and $\varphi_{1/v}$, for heterogeneous and homogeneous shields, is due entirely to the uncertainties in the cross sections for elastic scattering. In the heterogeneous shields, this error is twice that for the homogeneous shields; this result is attributed to an increase in the total thickness of the shield, and it leads to an increase in the factor by which the radiation is attenuated.

The difference between the roles played by elastic and inelastic scattering for neutron sources of the intermediate-neutron-converter type and of the $f + 1/E$ type is due to an enrichment of the spectrum of the intermediate-

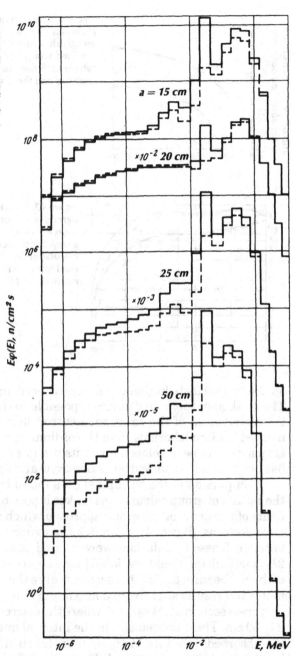

Figure 41. Energy flux density of the neutrons beyond iron barriers of various thicknesses *d*. Solid line, group calculation; dashed line, refined group calculation, with a calculation of the fraction of unscattered neutrons and of the first-collision sources by the method of subgroups.

neutron-converter source with neutrons of intermediate energies, which are slowed by elastic scattering; the $f + 1/E$ source has a harder spectrum, and the slowing in that case is determined to a large extent by inelastic scattering.

Let us examine the errors which stem from ignoring the spatial dependence of the group macroscopic interaction cross sections. The spatial dependence of the group-averaged cross sections, which is ignored in practical shielding calculations which use the group approximation, has the greatest effect

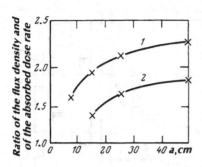

Figure 42. Ratio of the flux density (1) and of the neutron absorbed dose rate in tissue (2) behind Fe barriers in a group calculation to the corresponding characteristics in a refined group calculation, in which the method of subgroups is used to calculate the fraction of unscattered neutrons and the first-collision sources.

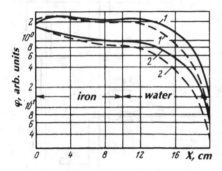

Figure 43. Spatial distribution of the neutron flux density in an iron-water shield for a unidirectional source (1) and a cosinusoidal source (2). Solid line, group calculation; dashed line, refined group calculation in which the method of subgroups is used to calculate the fraction of unscattered neutrons and the first-collision sources.

on the spatial distribution of the unscattered and singly scattered neutrons. The method of subgroups makes it possible to refine these distributions, like the distribution of the resultant neutron flux density. Since the subgroup method is more laborious than the ordinary group method, and special programs have to be developed for it, its use in practical shielding calculations has been limited to unscattered neutrons and first-collision sources.[159]

Comparison of the calculated data found by the subgroup method with the results of group calculations makes it possible to refine one of the components of the error of the group approach which stems from the values of the cross sections. Figure 41 compares the neutron energy spectra behind iron barriers. These calculations were carried out by the ROZ-9 program[102] (the $2P_N$ method) for a unidirectional plane source of neutrons with the spectrum of the B-2 beam of the BR-10 reactor. We see that the group approach overestimates the results slightly, to a maximum extent at an antiresonance in the iron cross section at 20–50 keV, where the discrepancy exceeds a factor of 2 for $d = 50$ cm. The discrepancies in the integral quantities (the flux density and the absorbed dose rate in tissue; Fig. 42) turn out to be quite large as the barrier thickness is increased. The same differences were found for an iron-water shield (10 cm Fe + 10 cm H_2O) bombarded by a unidirectional or cosinusoidal source[160] (Fig. 43). These examples illustrate the extent of the error in the ordinary group method for shielding calculations when the shield contains layers of iron, which has a clearly defined resonance structure in its cross sections.

The error of the group calculations can be reduced by choosing the group cross sections in a special way: by using the zeroth harmonic of the total

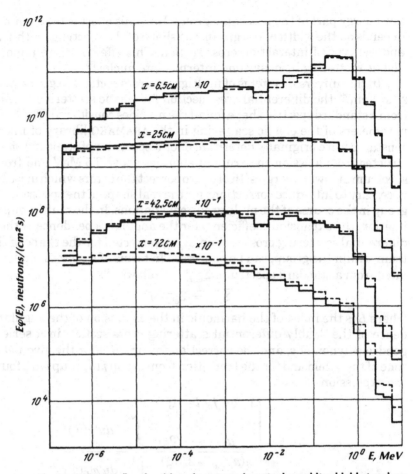

Figure 44. Energy flux densities of neutrons in a steel-graphite shield at various distances (x) from the source. Solid line, group calculation with a special choice of constants; dashed line, group calculation with a calculation by the subgroup method of the fraction of unscattered neutrons and of the first-collision sources.

"shielded" cross section in the calculation of the distribution of the unscattered neutrons and by using the "unblocked" scattering cross section in the calculation of the sources of singly scattered neutrons.[159] The situation is illustrated by Fig. 44, which shows the neutron energy flux densities in a heterogeneous steel-graphite arrangement (problem 9, Table V). The difference between the results in this case does not exceed 40% within an energy group.

We turn now to the error which results from the shape of the spectrum within a group and from the convolution of the multigroup constants. The multigroup constants of the doubly differential neutron scattering cross section depend strongly on the neutron spectrum within a group.[161] The reason why this cross section depends on the spectrum within a group is that at a group width considerably greater than the magnitude of the energy loss due to scattering the neutrons in a given group will come exclusively from the low-energy part of the preceding group. How this effect is taken into account depends on the width of the group, the shape of the spectrum within a group, and the type of interaction cross section. This effect is more important for elastic scattering by heavy and intermediate nuclei.[86]

In the universal systems of multigroup cross sections, e.g., the ARAMAKO-2F system,[86] the differential cross section for elastic scattering is calculated from a standard spectral shape for all the nuclides considered. The anisotropy parameters of the elastic scattering in the ARAMAKO library of microscopic constants were originally determined for a spectrum of standard shape, constructed from a fission spectrum at energies above 2.5 MeV and from a $1/E$ spectrum at lower energies. In order to correct the errors which arise here it is necessary to introduce corrections for the real shape of the spectrum within a group in the course of the multigroup calculations. In the multigroup approximation, the expansion coefficients for the angular dependence of the doubly differential scattering cross section, $\overline{\Sigma}_{el,L}^{j\rightarrow i}$, corrected for the shape of the spectrum within the group, can be expressed in terms of the cross sections calculated from a standard spectrum, $\Sigma_{el,L}^{j\rightarrow i}$, as follows[161]:

$$\overline{\Sigma}_{el,L}^{j\rightarrow i} = \Sigma_{el,L}^{j\rightarrow i} b_L^{j\rightarrow i}, \tag{3.39}$$

where L is the index of the harmonic in the expansion of the angular dependence of the doubly differential scattering cross section in a series in Legendre polynomials, and the correction factors $b_L^{j\rightarrow i}$ for the given geometric zone of the shield and for the transition from group j to group i are found from the expression

$$b_L^{j\rightarrow i} = \begin{cases} 1 \text{ for } j \neq i-1; \\ \dfrac{\varphi_L(u_{i-1}-2/3\xi)}{\varkappa(u_{i-1}-2/3\,\xi)} \dfrac{\displaystyle\int_{u_{j-1}}^{u_j} du'\varkappa(u')}{\displaystyle\int_{u_{j-1}}^{u_j} du'\varphi_L(u')}, \end{cases} \tag{3.40}$$

where $\varkappa(u)$ is the standard spectral shape adopted for the given library of cross sections, $\varphi_L(u)$ is the Lth harmonic of the neutron spectrum, averaged over the volume of the geometric zone of the shield under consideration, ξ is the average logarithmic energy loss, and u is the lethargy.

Using the results calculated on the fraction of neutrons with the uncorrected constants, $\Sigma_{el,L}^{j\rightarrow i}$, we can calculate the average energy distributions φ_L^i averaged over the geometric zones of interest. Analysis of various algorithms for approximating the neutron spectrum within a group[161,162] reveals the most convenient method to be a two-node linear interpolation of the multigroup neutron distribution.[161] The values of $b_L^{j\rightarrow i}$ calculated from (3.40) are used to correct the differential scattering cross section.

The role of this correction has been studied for typical shielding materials and layouts[159] by means of the ROZ-9 program[102] for calculations of the neutron field with the ARAMAKO-2F system of cross sections.[86,87] The neutron source was assumed to be a plane, unidirectional source with the spectrum of the beam in the B-2 installation of the BR-10 reactor, operated at a power of 1 MW (Ref. 163).

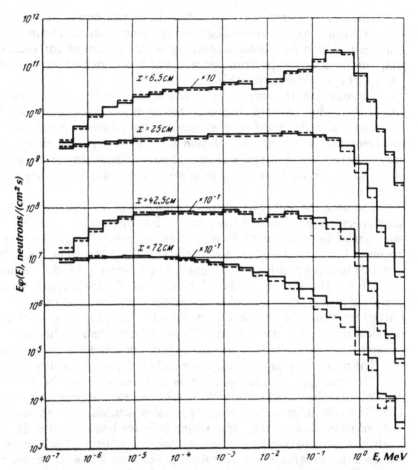

Figure 45. Energy flux density of neutrons in a steel-graphite shield at various distances (x) from the source. Solid lines, calculated with the constants before correction; dashed line, calculated with the constants corrected for the shape of the neutron spectrum within the groups.

Analysis of the neutron spectra beyond homogeneous barriers of iron and graphite with thicknesses up to 50 cm has shown[159] that the corrections to the differential scattering cross section do not cause discrepancies exceeding 10% in the spatial-energy distributions of the neutrons. The reason is the insignificant difference between the neutron spectra averaged over the volume of the zones and the standard spectrum adopted for the calculation of the original cross sections.

It is interesting to examine the effects of a correction of the cross sections for the shape of a spectrum within a group in thicker media, which cause a substantial deformation of the original neutron energy distribution. As an example of such a medium, a steel-graphite shield consisting of alternating layers of 1Kh18N9T steel and graphite was selected (calculation problem 9 in Table V; see also Fig. 7).

The maximum discrepancy between the spatial-energy distributions in Fig. 45 calculated from the constants before and after the correction reaches

100% in the energy range 1.0–10.5 MeV. In this particular shield layout, the ratio of the neutron dose rates absorbed in tissue calculated from the differential cross section for elastic scattering before and after correction for the shape of the neutron spectrum increases with distance from the source, from 1 to 1.3 at the exit from the shield.

These studies thus show that in calculations of the physical characteristics of the radiation fields in the shielding of nuclear-engineering installations, with significant differences between the neutron energy distributions and those assumed in the calculation of the group cross sections, it is necessary to correct the differential cross section for elastic scattering for the shape of the spectrum within a group in order to achieve the accuracy required in practice.

The need for highly accurate calculations of the neutron field, the complex energy dependence of the neutron interaction cross sections, and the substantial deformation of the neutron energy distributions in the shielding require a large number of groups. In the ARAMAKO-2F system,[86] e.g., 26 groups are used to solve problems pertinent to the shielding of fast reactors.

Most of the characteristics of the neutron field which are considered in the physics of reactor shielding are integral characteristics over the entire energy range or over wide energy groups. In this case, there is the possibility of calculating the neutron field for fewer groups than in the original multigroup partition, so that a significant amount of computer time can be saved.

The procedure for reducing the number of groups is called "collapsing."

The multigroup systems of cross sections such as the ARAMAKO system[86,87] and the OBRAZ system[164] must have a rather large number of groups (30–50). In the intermediate stage of physical studies, e.g., in the course of an optimization, an approximation which requires less computer time than the multigroup approximation—the so-called few-group approximation—should be used in programs for calculating neutron fields, as experience has shown.[85]

The few-group calculations are used in studies carried out to determine the relative changes in physical characteristics upon changes in the shielding parameters, e.g., the relative change in dose due to a change in the geometric dimensions of the zones or the nuclear concentrations of nuclides in the medium. The questions regarding the accuracy of the few-group approximation determine the reliability of the solution obtained with their help and also the extent of the additional multigroup studies which will be required to refine the solutions found in the few-group approximation.

When the number of groups is reduced, the boundary-value problem does not change in form; only the constants and the independent sources change. We will assume below that the boundaries of the groups of the original partition are the same as those of the partition after the convolution.

The homogeneous zones of the shielding are chosen in such a way that after the collapsing the importance functions of the neutrons within a group with respect to the functionals under consideration depend only weakly on the coordinates and the index of the group before the convolution. In this case it becomes possible to approximately determine effective few-group cross sections.[100,165]

A preliminary choice of group partition with fewer groups should be made on the basis of a sensitivity analysis itself based on multigroup calculations.[166,167] An energy-independent weight function can be used for the convolution in these studies.

As a criterion for choosing the few-group partition one could impose the condition that the various groups selected for the few-group partition must make equal contributions to the error of the physical characteristic on a consideration. If several functionals $\{R_k\}_k^k = 1$ are considered, the choice of a group partition with fewer groups can be based on the condition

$$\left|\frac{R'_k - R_k}{R_k}\right| \leqslant \left(\frac{\Delta R}{R}\right)_k, \quad k = 1,2,...,K, \tag{3.41}$$

where R'_k is the value of the functional of index k found in the calculation with the fewer groups, and $(\Delta R/R)_k$ is the accuracy required of the calculation of the given functional. Inequality (3.41) requires that the partition with fewer groups be chosen in such a way that the errors caused by the collapsing do not exceed the given errors.

For these purposes one could use the method of analyzing the sensitivity of a functional to the energy dependence of the interaction cross sections[166] which is embodied in the algorithm and program described in Ref. 167. In this algorithm, groups continue to be merged as long as the following inequality holds:

$$\left|\left(\frac{\delta R}{R}\right)_{i,k}\right| \leqslant \left(\frac{\Delta R}{R}\right)_{i,k}, \quad k = 1,2,...,K, \tag{3.42}$$

where $(\delta R/R)_{i,k}$ is the error in the functional in the enlarged group which is caused by the change in the group cross sections Σ^{ii} of the original group partition, given by

$$\left(\frac{\delta R}{R}\right)_{i,k} = \sum_{ii \in i} \sum_{\varkappa} p_k(\varkappa, ii) \, \sigma\Sigma_\varkappa^{ii}/\Sigma_\varkappa^{ii}. \tag{3.43}$$

The quantity $p_k(\varkappa, ii)$ in (3.43) is the relative sensitivity of functional R_k to a change in the interaction cross section of type $\varkappa$ in group ii, i.e., $\delta\Sigma_\varkappa^{ii}$.

The permissible error $(\Delta R/R)_{ii,k}$ can be found as the mean square deviation of the errors over the groups being combined which are caused by the original errors, $\Delta\Sigma_\varkappa^{ii}/\Sigma_\varkappa^{ii}$.

The procedure for a convolution using a given algorithm is illustrated by the example of the calculation of the equivalent dose rate of the neutrons from a source with an energy of 14 MeV behind a graphite barrier 2 m thick.[168] The functional and the sensitivity are calculated with the help of the programs of Refs. 56 and 169 and the constants of Refs. 82, 86, 165, and 170. As a result of the convolution, the number of groups is reduced by a factor of more than 2 in the energy range considered, above 0.8 MeV. The calculations carried out in this step by the methods of sensitivity analysis to find the group cross sections with a smaller number of groups yield information for a further collapsing of the multigroup constants into few-group cross sections.

Various approximations of the spatial-angular dependence of the solution of the transport equation in multigroup and few-group calculations for

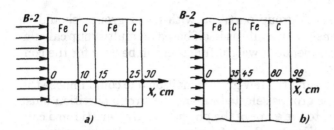

Figure 46. Shield layouts used to illustrate the convolution of multigroup constants into few-group constants.

neutron fields for two types of iron-graphite compositions (Fig. 46) were analyzed in Ref. 165. It was shown that the energy-angular distributions of neutrons integrated over the geometric zones considered must be used as a weight function for the convolution of the multigroup constants into few-group constants. In the calculation of the wave function required for the convolution of the cross sections, coarse difference approximations should be used in the multigroup calculation to save on computer time. The errors of the few-group calculations of the neutron field in shields consisting of extended homogeneous zones can be reduced through an additional partition of the geometric zones into finer zones.

4. Estimates of the errors and study of the sensitivity of calculations of radiation fields in models of real shields

Sensitivity analysis can have its most important practical application in problems of calculating functionals of radiation fields in the shields of actual nuclear-engineering installations. Conditions have been formulated for benchmark calculations of shields reflecting the basic features of real shields both in the Soviet Union (calculation model 12 in Table V) and elsewhere (calculation models 10 and 11 in Table V). The purposes of these projects were to estimate the errors in the basic functionals of the radiation fields, to study the sensitivity of these functionals to the interaction cross sections, and to test various libraries of cross sections.

Let us examine the results calculated for the radiation fields for the three shield models mentioned above, for a pressurized water reactor, a fast breeder, and a fast-neutron reactor.

Model for the shield of a pressurized water reactor

Using the data of Refs. 120 and 122–125 as examples, let us examine the characteristics of the radiation fields in a model shield for a pressurized water reactor (calculation model 10 in Table V; Fig. 7).

Table XXVIII shows values of the functionals of the radiation flux density, R_k, in a mockup of the shield of a reactor of this type: the rate of radiation

Table XXVIII. Characteristics of the fields of neutrons and secondary radiation in a mockup shield for a pressurized water reactor.

Functional R_k		Reference		
		[120]	[122]	[123]
RD, displacements/(atom s)		1.39×10^{-12}	1.975×10^{-12}	6.9×10^{-13}
P_{equ}, cSv/h [rem/h]	Neutrons	4.42×10^{-3}	1.1×10^{-3}	3.7×10^{-3}
	γ radiation	10.4×10^{-3}	40.7×10^{-3}	10.0×10^{-3}

damage at the inner surface of the reactor vessel due to neutrons (RD) and the equivalent dose rate at the outer surface of the concrete shield (P_{equ}).

The discrepancies observed in the results calculated by different investigators exceed the permissible errors in the calculation of the characteristics P_{equ} and the radiation damage (RD; Table III). For P_{equ} of both the neutrons and the secondary γ radiation, for example, there are discrepancies of up to a factor of 4, while the discrepancies in the calculations of the radiation damage exceed a factor of 2. It follows from a comparison of the data of Refs. 120, 122, and 123 that the discrepancies in the calculations of the heat evolution (HE) in a concrete shield exceed 30%. These discrepancies appear to be due primarily to discrepancies in the interaction cross sections used for the calculations on the radiation transport and in the detector response functions for the functionals considered.[171]

The relative sensitivities of the radiation damage, the heat evolution, and P_{equ} to the total cross sections of the elements in the shielding are listed in Table XXIX. The relative sensitivities of the radiation damage to the cross sections of the individual nuclides agree well for the studies considered. An exceptional case is the significant difference in the relative sensitivity of the radiation damage to the hydrogen cross section in Ref. 123.

It follows from an analysis of the relative sensitivities of this functional to the neutron interaction cross sections that the largest values of the relative sensitivity for the radiation damage are observed for H, O, and Fe. In terms of the relative sensitivity of this characteristic to the total cross section, the elements Cr, [238]U, Mn, Ni, and Zr can be put in a secondary group. All the other elements have insignificant relative sensitivities, so that we can ignore the errors in the estimates of their cross sections.

The largest relative sensitivities of the heat evolution to the interaction cross sections are seen for H, Fe, O (as for the radiation damage), and also [238]U.

The relative sensitivities of P_{equ} to the γ interaction cross sections in Refs. 120 and 124 are very nearly equal. The largest relative sensitivities of P_{equ} to the interaction cross sections are seen for O, Fe, H, Si, and Ca. According to the data of Ref. 124, if we adopt the total relative sensitivity of P_{equ} to the neutron cross sections to be -3.13, and if we take into account the maximum error of 30% in the total cross section, we find the error in P_{equ} to be 100% by a linear perturbation theory.

Assuming that the errors in the estimated cross sections of the most important elements are approximately the same, we can draw the following

Table XXIX. Relative sensitivities of the functionals of the radiation fields in the mockup of the shield of the pressurized water reactor to the total cross sections of the elements and isotopes.

R_k	RD			HE			P_{equ}			
				[120]		[122]	[120]		[124]	
Element or isotope	[120]	[122]	[123]	Neutrons	γ radiation	Total	Neutrons	γ radiation	Neutrons	γ radiation
H	−5.02	−5.67	−0.96	−2.71	−0.18	−1.25	−2.71	−0.13	−2.31	−0.06
O	−1.66	−2.00	−1.89	−0.39	−0.89	−0.35	−0.39	−4.66	−3.13	−4.94
Fe	−1.50	−1.60	−1.51	−0.04	−5.51	−0.68	−0.04	−2.90	−2.33	−2.83
Zr	−0.12	−0.11	−0.13	−0.07	−0.02	−0.03	−0.07	−0.004	−0.10	−0.01
235U	−0.01		−0.01	−0.02 ⎱	−0.12		−0.02 ⎱	−0.02	+0.01	−0.00
238U	−0.30	−0.30	−0.31	−0.23 ⎰		−0.14	−0.23 ⎰		−0.31	−0.06
Cr	−0.36	−0.46	−0.42	−0.06	−0.30	−0.13	−0.06	−0.15	−0.31	−0.15
Ni	−0.21	−0.23	−0.21	−0.04	−0.18	−0.08	−0.04	−0.10	−0.13	−0.10
Mn	−0.27		−0.10	−0.01			−0.01	−0.01	−0.02	−0.01
Al					−0.005			−0.38	−0.16	−0.41
Si		∼10⁻⁴		∼10⁻⁴	−0.02		∼10⁻⁴	−1.60	−0.57	−1.71
Ca					−0.03			−1.73	−0.62	−1.86
a	−9.46	−10.37	−5.54	−3.57	−7.29	−2.66	−3.57	−11.70	−9.99	−12.16

ᵃ All elements and isotopes.

conclusion from these relative sensitivities (Table XXIX): The errors in the cross sections of O, Fe, and H have the greatest effect on the accuracy of the calculations of the radiation damage and the heat evolution; in addition, the cross sections of Si and Ca are among the most important for P_{equ}.

From an analysis of the relative sensitivities of the radiation damage to the total cross sections of the individual elements, carried out in Ref. 125, it was concluded that in order to attain a given error of 20% in the calculations of the radiation damage in the shielding model adopted it was necessary to know the total cross sections over the energy range from 4 to 10 MeV within errors no lower than 0.5% for H; 1.5% for O; 2% for Fe; 10% for [238]U, Cr, and Ni; and 30% for Zr.

To illustrate the estimates of the errors in the calculations of the functionals as functions of the approximation (P_L) of the differential scattering cross section in the group approximation, we consider the elements having the maximum sensitivities. Table XXX lists the changes in percent in the characteristics of the radiation fields considered upon a switch from the P_3 approximation to P_2, P_1, and P_0, both for individual elements and for all elements simultaneously. Analysis of the data in Table XXX leads to the conclusion that the P_0 and P_1 approximations are unacceptable for calculations of the radiation damage within an error of 10%–15%. The P_3 approximation in the differential scattering cross section leads to the specified accuracy in the calculation of the radiation damage. In the absence of data on the differential scattering cross section in the P_3 approximation, we can use data on the angular dependence of the differential cross section for scattering in the P_2 approximation.

The data found from the studies cited agree in general. There are significant discrepancies in the results calculated for hydrogen, apparently because of the particular features of the method used in Ref. 123 to calculate the radiation field.

Analysis of the data in Table XXX shows that in order to achieve an error of 30% in the calculation of the heat evolution it is necessary to use at least the P_2 approximation of the angular dependence of the differential scattering cross section. The P_1 approximation is sufficient to achieve a 100% error in the calculation of P_{equ} for secondary γ radiation.

In calculations on the neutron field and the field of secondary γ radiation, if it is necessary to calculate the set of these functionals, the most accurate approximation P_L should be used.

To analyze the importance of the individual energy ranges of the neutrons and the secondary γ radiation in calculations of the given characteristics of radiation fields, one should use information on the energy dependence of the relative sensitivity to the total cross sections.[120,122,124,125] Figures 47 and 48 illustrate the situation with examples of the energy dependence of the relative sensitivity of P_{equ} for secondary γ radiation at the exit from concrete to the total cross sections for the interactions of neutrons or secondary γ radiation.[124]

Analysis of these distributions reveals that P_{equ} of the secondary γ radiation is highly sensitive to the interaction cross sections of fast neutrons with energies above 2 MeV. The relative sensitivity is observed to remain constant over the neutron energy range from 0.5 eV to 100 keV (Fig. 48). The high sensitivity for thermal neutrons is noteworthy.

In the energy dependence of the relative sensitivity of P_{equ} of the secondary γ radiation to the neutron interaction cross sections for elements which do not have a resonance structure in their cross sections, we see minima and maxima corresponding to the energy dependence of the cross sections for the elements which do have a resonance structure. In the energy dependence of the relative sensitivity of P_{equ} of the secondary γ radiation to the H cross section (Fig. 47, for example), we see maxima and minima corresponding to the resonance structures of O, Fe, Si, and [238]U.

The relative sensitivity of P_{equ} of the secondary γ radiation to the total cross sections for the interaction of γ radiation with matter decreases with decreasing energy.

There is considerable interest in the relative contributions to a given functional R_k from broad energy intervals. Information of this sort can be used to generate objective estimates of the relative contributions to the physical characteristics of interest from specific energy ranges. The relative contributions to the radiation damage found in Ref. 123 for energies above a given level show that neutrons with energies above 1.35 MeV contribute 65% of the radiation damage, while neutrons with energies above 0.82 MeV contribute 78%. We can thus conclude that fast neutrons play a governing role in a calculation of the radiation damage.

The relative sensitivities of a given functional to the total cross sections, integrated over certain energy ranges, do not give us comprehensive information about absolute contributions. They show only the effect of a change in the cross section in this region on the particular functional. The integral sensitivities which are found thus indirectly point out the energy regions which are important for a subsequent calculation of absolute contributions.

To illustrate these relative sensitivities integrated over certain energy

Table XXX. The changes $\Delta R_i/R_i$ (%) in the characteristics considered according to various approximations (P_l) of the differential scattering cross section.

Element or isotope	RD [120]			RD [122]			RD [123]			HE [120] Neutrons			HE [120] γ radiation			HE [123] Common representation			P_{equ} [120] Neutrons			P_{equ} [120] γ radiation		
	P_0	P_1	P_2	P_0	P_1	P_2	P_0	P_1	P_2	P_0	P_1	P_2	P_0	P_1	P_2	P_0	P_1	P_2	P_0	P_1	P_2	P_0	P_1	P_2
H	−99	−22	−1.3	−116	−30	−4.9	−12	−2	0	−91	−2.5	−0.02	−2.5	−0.8	−0.1	−31	−2	0	−91	−2.5	−0.02	−1.2	−0.3	−0.02
O	−55	−8.8	−0.5	−47	−12	−2.2	−42	−8	−1.0	−17	−0.5	0	−4.2	−3.4	−1.5	−112	−29	−4	−17	−0.5	0	−61	−14	−1.4
Fe	−57	−12	−2.6	−69	−21	−6.1	−58	−15	−3.0	−12	−0.03	−0.02	−91	−29	−3.8	−114	−30	−5	−12	−0.03	−0.03	−13	−7.9	−2.9
238U										−3.7	−0.20	−0.02	−0.05	−0.04	−0.03	−5	+1	0						
Si																						−20	−4.7	−0.5
Ca																						−20	−4.7	−0.5
a	−250	−48	−4.4	−260	−70	−14.5	−115	−29	−4	−118	−2.8	−0.03	−98	−35	−6.8	−356	−85	−10	−118	−2.8	−0.03	−122	−33	−5.8

a All elements and isotopes.

Table XXXI. Characteristics of the neutron fields in the shielding of a fast breeder reactor.

r, cm	$\varphi(r)$ (n/cm² s) [127]	[127]/[172]	$\Phi(r)$ (n/cm² s) [127]	[127]/[172]	$\varphi_i(r)$ (atom⁻¹ s⁻¹) [127]	[127]/[172]	$RD(r)$ (atom⁻¹ s⁻¹) [127]	[127]/[172]	$AL(r)$ (atom⁻¹ s⁻¹) [127]	[127]/[172]	$FR(r)$ (atom⁻¹ s⁻¹) [127]	[127]/[172]	$AR(r)$ (atom⁻¹ s⁻¹) [127]	[127]/[172]
262	2.26+13	0.88	2.76+11	1.4	3.55+12	1.3	1.02+15	0.6	2.13+11	0.80	2.99+14	1.5	1.22+14	2.0
292	4.71+12	1.18	8.24+10	1.7	5.80+11	2.5	1.57+14	1.0	5.73+10	1.0	7.78+13	1.6	2.88+13	1.4
322	8.50+11	1.6	1.64+10	1.0	8.76+10	1.4	2.39+13	1.4	1.12+10	1.0	1.50+13	1.5	5.43+12	1.3
352	1.43+11	2.3	2.94+9	1.5	1.27+10	7.1	3.53+12	1.9	1.99+9	0.80	2.65+12	1.2	9.53+11	1.2
382	2.28+10	3.3	4.90+8	2.5	1.79+9	15	5.09+11	3.4	3.29+8	0.90	4.37+11	1.5	1.56+11	1.7
415	2.68+9	3.0	7.13+7	3.6	1.79+8	20	5.32+10	4.1	4.57+7	1.5	5.79+10	1.9	1.95+10	2.5
418.5	2.32+9	3.9	6.99+7	3.9	1.44+8	24	4.46+10	4.5	4.28+7	1.7	5.40+10	1.9	1.73+10	2.4
498.5	5.55+9	4.6	3.83+7	5.8	6.97+8	77	3.30+9	13	2.17+7	2.7	2.41+10	2.0	5.95+9	3.3
578.5	1.19+8	4.8	1.25+7	7.0	3.51+5	350	3.46+8	10	6.92+6	3.5	7.06+9	2.0	1.25+9	4.2
666.5	1.90+7	5.4	2.69+6	8.4	1.34+4		4.90+7	6.1	1.47+6	3.9	1.41+9	2.6	1.80+8	4.5
738.5	3.77+6	5.4	6.39+5	11	9.37+2		1.07+7	5.1	3.48+5	5.0	3.23+8	3.2	3.44+7	3.4
818.5	5.65+5	5.1	1.11+5	12	4.95+1		1.83+6	7.0	6.04+4	5.5	5.51+7	3.4	5.19+6	3.5
914.5	5.15+4	6.2	1.06+4	13	1.55+0		1.73+5	9.5	5.76+3	6.4	5.17+6	3.1	4.62+5	4.2
917.5	4.55+4	6.5	9.15+3	18	1.39+0		1.48+5	12	4.95+3	7.0	4.43+6	3.0	4.00+5	4.0
932.5	1.63+4	4.4	2.83+3	19	7.07−1		4.53+4		1.54+3	4.4	1.34+6	3.4	1.28+5	4.6
947	5.83+3	4.9	9.40+2	9.4	3.46−1		1.50+4		5.13+2	2.0	4.39+5	3.1	4.29+4	3.6
965	1.67+3	4.2	2.77+2	7.8	1.40−1		4.44+3		1.51+2	1.9	1.29+5	1.7	1.24+4	3.1

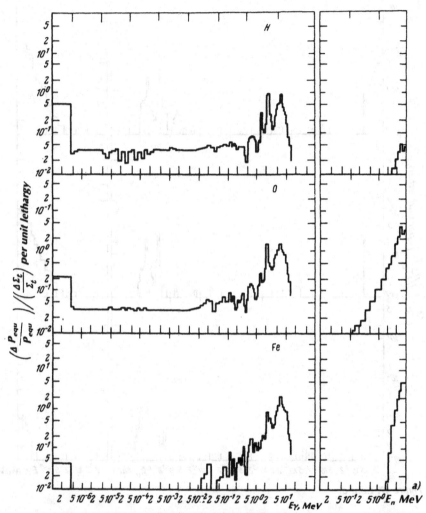

Figure 47. Energy dependence of the relative sensitivity of the dose equivalent rate of secondary γ radiation at the exit from a concrete shield, P_{equ}, to the total cross section for the interaction of (a) neutrons and (b) γ radiation with H, O, Fe, Si, Ca, and ^{138}U nuclei.

ranges, we might cite the data of Ref. 122 on heat evolution. Neutrons with energies above 100 eV are responsible for no more than 10% of the integral sensitivity. Almost 80% of the sensitivity corresponds to thermal neutrons. These results are evidence that the secondary γ radiation from radiative capture makes the leading contribution to this functional.

The energy dependence of the relative sensitivity is used to determine those errors in the characteristics of radiation fields in shielding which stem from errors in the interaction constants. As an example, we can cite the results of a calculation of the errors in radiation damage due to uncertainties in

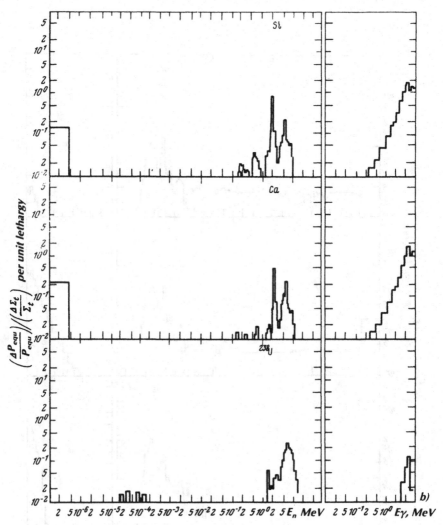

Figure 47. Cont'd.

the Fe and O group cross sections.[120] The uncorrelated errors in the radiation damage which stem from errors in the Fe, O, Fe + O, and all-element group cross sections are 3.30%, 3.65%, 4.93%, and 6.56%, respectively. The maximum errors in the radiation damage due to these cross sections are roughly 2.3 times the uncorrelated errors.

Model for the shield of a fast breeder reactor

Let us examine the basic results found in an analysis of the characteristics of the radiation fields calculated for a model for the lateral shielding of a fast breeder reactor[125–127,172] (calculation model 11 in Table V; Fig. 7). Table XXXI shows data for comparing the results of the calculations of the basic characteristics of the neutron field in the shielding. The following characteristics R_k

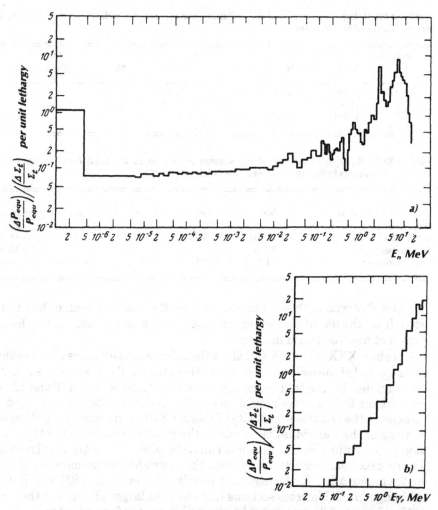

Figure 48. Energy dependence of the relative sensitivity of the equivalent dose rate of secondary γ radiation at the exit from concrete, P_{equ}, summed over all the nuclides in the shielding, to the total cross sections for the interaction of (a) neutrons and (b) γ radiation.

were selected: the rate of radiation damage (RD) in Fe at the exit from the radial Fe-Na shield; activation of the Na of the secondary loop in the heat exchanger (AL); the total neutron flux φ; the heat flux φ_t; the flux of fast neutrons with energies above 0.111 MeV, φ_F; the rate of the ^{235}U fission reaction (FR); and the rate of the ^{59}Co activation (AR). This comparison of the results of Refs. 127 and 172 for various distances (R) from the center of the core shows that the discrepancy in terms of the radiation damage is more than a factor of 4. The discrepancies in the results on the sodium activation rate in the zone of the heat exchanger range from a factor of 2 to a factor of 7. For several characteristics, the discrepancies in the results calculated from different programs can reach a factor of 10. The results of Ref. 127 are systematically higher than the results reported in Ref. 172. The significant discre-

Table XXXII. Relative sensitivity of the radiation damage and the sodium activation to the total cross sections of the elements.

Functional R_k	Study	Fe	Na	Cr	Ni	Total for the given elements
RD	[127]	− 4.74	− 2.95	− 1.63	− 1.37	− 10.69
AR	[126]	− 8.18	− 20.81	− 2.63	− 4.71	− 36.34
	[127]	− 5.40	− 14.89	− 1.79	− 1.56	− 23.64

Table XXXIII. Relative sensitivity of the sodium activation to the total cross sections of the individual elements in various regions of the shield (Ref. 126).

Shield region	Na	Fe	Cr	Ni	All elements
Lateral shield	− 4.30	− 6.06	− 2.30	− 3.75	− 16.41
Sodium basin	− 15.18				− 15.18
Heat exchanger	− 1.17	− 2.12	− 0.33	− 0.96	− 4.59

pancies observed in these characteristics of the neutron field in the shielding stem from the use of very different data on the group constants for the interaction of neutrons with matter.[171]

Tables XXXII and XXXIII show the relative sensitivities of the radiation damage and of the sodium activation rate to the total cross sections of individual nuclides. We see that, although the calculations in Refs. 126 and 127 were carried out by the same programs (SWANLAKE and ANISN), there are discrepancies in the relative sensitivity (Table XXXII). Correspondingly, the contributions of the individual elements to the relative sensitivity to the interaction cross sections of the elements can be broken up into groups. The Na and Fe cross sections are very important. For the radiation damage, for example, a 1% change $\Delta\Sigma/\Sigma$ for Fe leads to a nearly 5% change in $\Delta RD/RD$ (Ref. 127). A change in the Na cross sections leads to even larger changes in the sodium activation rate. Chromium and nickel fall in a second group in terms of importance. All the other elements are of less importance for the characteristics considered here, and we will accordingly ignore the errors in their cross sections.

Estimates of the relative sensitivity to the various angular approximations of the differential scattering cross section are compared in Table XXXIV. These results show that the P_1 approximation is sufficient for calculations on both the radiation damage and the sodium activation. The difference between the values found for these characteristics in this approximation and in the P_3 approximation will not exceed 2%.

Figure 49 shows the energy dependence of the relative sensitivity of the sodium activation rate to the total cross sections in individual zones. For the Fe + Na lateral shield, the energy region above 100 keV is the most important. In the sodium basin we observe two energy regions which determine the contribution of neutrons to this functional. The contribution to the sodium activation rate is due primarily to neutrons with energies below 1 keV. In the heat exchanger, neutrons with energies below 100 eV play a leading role.

Table XXXIV. Relative change $\Delta R_k / R_k$ (%) in specified characteristics of the neutron field in a model of the shield of a fast breeder reactor for various approximations P_L of the differential scattering cross section.

Element	Sodium activation-AK		Radiation damage-PH	
	$P_0(P_1)^a$ [126]	$P_1(P_3)$[127]	$P_0(P_1)$[126]	$P_1(P_3)$[127]
Na	− 66	− 0.30	− 67.1	− 0.25
Fe	− 67	− 0.83	− 16.9	− 0.92
Cr	− 29	− 0.31	− 5.17	− 0.34
Ni	− 23	− 0.14	− 2.20	− 0.15
All elements	− 185	− 1.58	− 91.37	− 1.66

^{a} $P_L (P_m)$ means the switch from the P_m to the P_L aprproximation.

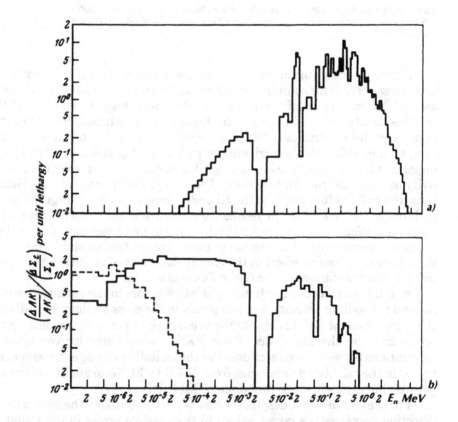

Figure 49. Energy dependence of the relative sensitivity of the activation (AR) of the sodium in the secondary loop to the total cross section for neutron interactions. (a) At the lateral Fe + Na shield; (b) at the sodium basin (solid line) and at the heat exchanger (dashed line).

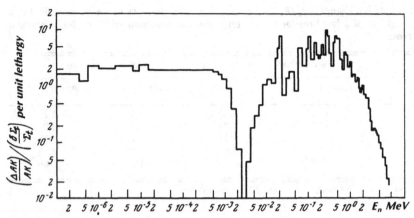

Figure 50. Energy dependence of the relative sensitivity of the activation (AR) of sodium in the secondary loop, summed over all regions of the shield, to the total cross section.

Figure 50 shows the energy dependence of the relative sensitivity to the total cross section for sodium activation, summed over all zones and all elements.[126] Comparison of these results with the energy dependence of the relative sensitivity for sodium (Fig. 51) leads to the conclusion that the contributions to the relative sensitivity for energies above and below 3 keV are comparable, although the high-energy part of the spectrum is slightly less important for Na. For Fe there is a significant decrease in the relative sensitivity at energies from 10 to 500 eV. The energy dependence of the relative sensitivity of this functional to the Ni and Cr cross sections is largely similar to the dependence for Fe. The resonance structure in the elements present in this shield affects the relative sensitivities to the cross sections of more than the given element; e.g., the resonance structures in the Na and Ni cross sections have a noticeable effect in the energy dependence of the relative sensitivity of the sodium activation to the Fe cross section.

Finally, we consider the limiting relative errors in the radiation shielding and the sodium activation due to errors in the cross sections according to the calculations of Ref. 127. As for the pressurized water reactor, data on the errors in these characteristics (Table XXXV) were found for two types of correlations between cross sections. For the radiation damage, the errors expected in the calculated data range from 21.7% to 30.9%, or slightly above the permissible errors in these functionals (Table III).

For the fast breeder, despite the shorter distances from the source to the detection point, we see larger values in the possible errors in the radiation damage than for the pressurized water reactor. Consequently, the requirements on the accuracy of the constants may be more severe in calculations of the radiation damage for the shielding in fast reactors than for thermal reactors. For the sodium activation, the possible errors in the calculated results may reach 70%.

Analysis of the calculations shows that the discrepancies in the results of these studies are greater than the errors which would be expected because of the errors in the cross sections.

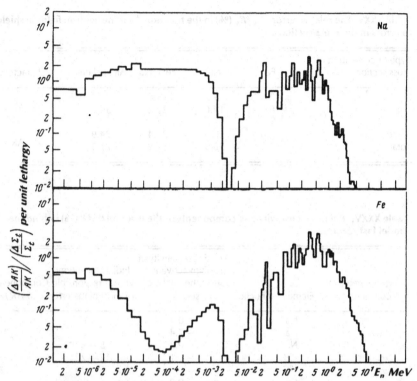

Figure 51. Energy dependence of the relative sensitivity of the activation (AR) of the sodium in the secondary loop to the total cross sections of Na, Fe, Ni, and Cr.

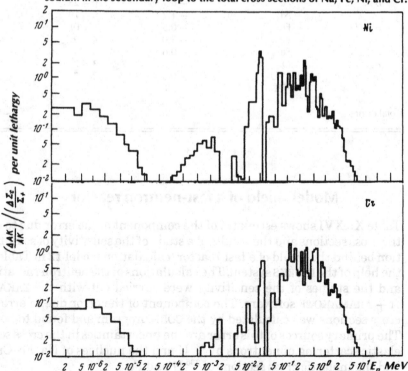

Table XXXV. The relative error $\delta R_k / R_k$ (%) in the functionals of the neutron fields in shielding due to errors in the cross sections.

Type of correlation of cross sections	Fe	Na	Fe + Na	All nuclides	Characteristic
None	12.3	9.7	15.6	21.7	RD
Total	16.8	14.6	22.2	30.9	
None	13.6	16.6	21.4	24.9	AR
Total	22.5	56.9	61.2	71.3	

Table XXXVI. Relative sensitivity and components of the error $\delta AR/AR$ (%) behind the shield of a model fast reactor.

Type of neutron constant, σ_n	Element of nuclide	Relative sensitivity of sodium activation rate to the total cross section	Indices of energy groups responsible for most of the error	Components of error $\delta \overline{AR}/AR$, %
σ_c	Na	−0.1	>19	1
σ_{el}	Na	−2.4		25
σ_{in}	Na	−0.2	5−6	3
σ_c	^{10}B	−1.1	17−22	1
σ_{el}	^{12}C	−9.5	6−20	8
σ_c	Cr	−0.2	>20	1
σ_{el}	Cr	−1.9	5−10	19
σ_c	Ni	−0.1	20−23	1
σ_{el}	Ni	−1.2	6−10	12
σ_c	Fe	−0.5	>19	1
σ_{el}	Fe	−5.0	4−9	28
σ_{in}	Fe	−0.6	3−5	8
σ_{el}	O	−1.3	3−7	4
σ_{el}	^{238}U	−0.5	2−7	8
σ_{in}	^{238}U	−1.0	3−7	12
Total error				~50%

Model shield of a fast-neutron reactor

Table XXXVI shows estimates of the component of the error due to errors in the cross sections and the results of a study of the sensitivity of sodium activation behind the shield of a fast reactor (calculation model 12 in Table V) with the help of the INDÉKS system. The calculations of the neutron radiation fields and the studies of the sensitivity were carried out with the ZAKAT + ROZ-11 + ARAMAKO-2F software. The component of the error due to errors in the cross sections was calculated by the CORE program and found to be ±50%. The primary sources of this error are the uncertainties in the cross section for elastic scattering of neutrons by ^{238}U and the nuclides of Na, Fe, Cr, and Ni and the inelastic cross sections of ^{235}U and Fe.

Figure 52. The relative readings R'/R of a detector which measures neutrons with energies below 160 keV vs the relative value σ'/σ of the total cross section for interaction of neutrons with Na at energies near 297 keV. Solid line, exact calculation by the direct-substitution method; dot-dashed line, linear perturbation theory; dashed line, second-order perturbation theory; filled circles, SACF method under the assumption of a linear dependence of p on the interaction cross section.

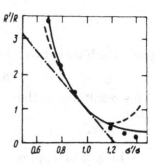

5. Estimate of nonlinear effects in sensitivity analysis

When expression (1.17) is used to calculate the relative variation in a functional, $\Delta R/R$, and Eq. (1.18) is used to estimate the relative error of a functional, $\delta \overline{R}/R$, there is the question of whether linear perturbation theory is applicable for the specific perturbations $\Delta X_1/X_1, \ldots, \Delta X_N/X_N$ and for the specific uncertainties in the input parameters, $\delta \overline{X}_1/X_1, \ldots, \delta \overline{X}_N/X_N$. The accuracy of calculations of these properties can be improved by moving up to a higher-order (e.g., second-order) perturbation theory.[173]

Figure 52 shows the relative value of the functional, $R'/R = (R + \Delta R)/R$, as a function of the relative value of the parameter, $\sigma'/\sigma = (\sigma + \Delta\sigma)/\sigma$, according to exact calculations, according to linear perturbation theory,[84] and according to second-order perturbation theory.[173] The functionals R which were considered were the readings of a detector which detected neutrons at a constant efficiency at energies below 160 keV after they had passed through 260 cm of sodium, for a unidirectional neutron source with an energy of 360 keV. The parameter which was varied, σ', was the total group cross section for the interaction of neutrons with sodium, which has a minimum in its cross section at 297 keV. This problem is typical in research on the deep penetration of radiation, where the resonance structure of the cross sections is important: The actual errors in the cross sections lead to large changes in the value of the functional under consideration. Analysis of the data in Fig. 52 shows that second-order perturbation theory is considerably better than linear perturbation theory; it describes the dependence of R'/R on σ'/σ, but it does not take into account the asymmetry of R'/R. The use of a higher-order perturbation theory will obviously improve the accuracy of the calculation of R'/R, but the efficiency of the calculations worsens with increasing order.

There is another way to improve the accuracy of the calculations: Expressions (1.21) and (1.22) can be used to calculate $\Delta R/R$ and $\delta\overline{R}/R$, respectively. Those expressions assume knowledge of the dependence of the relative sensitivity $p_{Rn}(X_1, \ldots, X_N)$ on the parameter X_n. Before the calculations are carried out, one can specify a first approximation of the dependence of p_{Rn} on the parameters X_n in the operator L. Using $R = \langle \varphi, L^* \varphi^* \rangle$, and substituting expression (1.27) into expression (1.29), we find

$$p_{Rn}(X_1,...,X_N) = \frac{\langle \varphi, X_n L_n^* \varphi^* \rangle}{\langle \varphi, L^* \varphi^* \rangle}, \tag{3.44}$$

from which we in turn find the important result

$$p_{Rn}(X_1,...,X_n = 0,...,X_N) = 0. \tag{3.45}$$

This result means that p_{Rn} depends on the parameter X_n in the following way:

$$p_{Rn}(X_1,...,X_N) = X_n f_n(X_1,...,X_N). \tag{3.46}$$

In a first approximation the relative sensitivity may therefore be thought of as a linear function of X_n. The function f_n is a smooth function of the parameters $X_1,...,X_N$, since expression (3.44) is a fractionally bilinear functional of φ and φ^*.

The points in Fig. 52 are values of R'/R calculated from Eq. (1.20) under the assumption of a linear dependence of p_{Rn} on X_n (i.e., for f_n = const). The procedure of using (1.20) for calculations has been labeled the method of "semianalytic calculations of a functional" (SACF).[74] We see that this method not only provides results which agree better than those of second-order perturbation theory with the exact results (Fig. 52) but also is equivalent to linear perturbation theory in terms of the volume of calculations. It should be noted that a significant dependence of a functional on an input parameter is typical when, for example, the parameter describes structural features (resonances) in the energy dependence of the cross section for the interaction of radiation with matter. Even in this case, however, the linear approximation of p_{Rn} as a function of X_n works well. For other parameters, we do not have this strong dependence. We thus have confidence that the linear approximation of p_{Rn} as a function of X_n will be useful in practical studies on nonlinear effects when the range of a parameter is narrow, as in the case of uncertainties or errors in interaction cross sections.

The construction of more accurate approximation formulas for $p_{Rn}(X_1,...,X_N)$ will make it possible to reconstruct the analytic dependence of R on $X_1,...,X_N$ over broad ranges of the input parameters.[74] This facility is important, for example, in the construction of approximate formulas for optimizing shielding and for interpreting the results of calculations carried out with various libraries of cross sections. Finally, we note that when $\Delta R/R$ is a nonlinear function of $\Delta X_n/X_n$ (Fig. 52) the following relation does not hold:

$$\overline{R(X_1,...,X_N)} = R(\overline{X}_1,...,\overline{X}_N). \tag{3.47}$$

Here the superior bar means a mathematical average. Both in the Soviet Union and elsewhere, relation (3.47) is used in calculations. It is obtained when R is a linear function of $X_1,...,X_N$ or when the distribution function of $X_1,...,X_N$ is a δ-function distribution. In general, a correction

$$\omega_R = \overline{R(X_1,...,X_N)}/R(\overline{X}_1,...,\overline{X}_N) \neq 1 \tag{3.48}$$

should be made to the calculations.

The deviation of ω_R from 1 increases as the dependence of $\Delta R/R$ on $\Delta X_n/X_n$ becomes progressively more nonlinear. The correction ω_R will apparently be important in deep-penetration problems, and it must be taken into account in comparisons of calculated and experimental data.

Chapter 4
Experiments in sensitivity analysis and in the determination of calculation errors

1. Benchmark experiments and sensitivity analysis

The balance between experiments and calculations has been constantly shifting in the favor of the latter,[49,174,175] primarily because of the steady progress in computer technology and the trend toward more complicated and more expensive experiments.

Any experimental study (Fig. 1) can be classified as either a reference experiment, whose primary purpose is to test calculation methods and the libraries of cross sections for the interaction of radiation with matter, or a parametric experiment, carried out to solve one of the few problems which cannot at present be solved satisfactorily accurately by calculations.

Among the reference experiments, those which dominate the literature are benchmark (the corresponding Russian word is bazovyĭ) experiments carried out under clean, standard conditions which are as simple as possible (in a simple one-dimensional geometry or in a multidimensional geometry), with a source, a shield, and a detector whose characteristics are well known. In benchmark experiments on the propagation of radiations the uncertainties which stem from the theoretical treatment of the parameters of the source and the geometry of the system are small in comparison with the measurement errors at the given depth in the shielding.[6,176]

This criterion for a benchmark experiment makes it possible to select from the abundance of experimental data in the literature those experiments which can be classified as benchmark experiments. Information on these reference benchmark experiments for a clean geometry can be used both to test and to correct (adjust) cross sections for the interaction of radiations with matter in the course of a design project. Progress in research on the sensitivity of calculated results to the input parameters has now made it possible to examine these two directions in the use of integral benchmark information (on various functionals of the radiation field which are independent of the angle and of the energy) from a strictly quantitative point of view. (In general, both integral and differential characteristics of a field can be determined in benchmark experiments.)

Experimental apparatus (Table XXXVII)

Plans for benchmark experiments and sensitivity studies were worked out at an international meeting[6] in 1974. Since then, four special international meetings have been held on this question,[3,177,178] and the first results of benchmark experiments have been published in standard format. At the first meeting,[6] various types of experimental apparatus available for use in an overall program of benchmark studies were analyzed.[6,171,179,180]

The development of a program of benchmark studies has also received considerable attention in the Soviet Union. These questions have been discussed extensively at nationwide scientific conferences on shielding against ionizing radiation in nuclear-engineering installations.[9-11] These conferences generated goals for future experiments: (a) to compile and evaluate data from benchmark experiments carried out by various groups in the Soviet Union and elsewhere for subsequent use for testing calculation methods and programs and for correcting the constants used to describe the interaction with matter; (b) to carry out new reference experiments under conditions approximating practical problems in shielding, including experiments in multidimensional geometry.

Apparatus with a fission-neutron converter, such as ASPIS in Winfrith, TRIGA in Casacia, EURACOS II in Pavia, and KND in Obninsk, is useful for benchmark studies. In addition, installations at fast reactors can be used: the BR-10, HARMONIE, TAPIRO, and YAYOI. These installations have various radiation and geometric characteristics (energy spectrum, source strength, system geometry). Variations of these parameters make it possible to carry out experiments with quite different relative sensitivity functions. Experiments of this sort should be carried out in a clean geometry of the source and the shield, with ^{252}Cf fission sources and 14-MeV neutrons from the T(d,n)He reaction. Furthermore, benchmark experiments should be carried out in such a way that they can be interpreted directly through the use of the design calculation programs, operated under ordinary conditions.

Benchmark experiments should begin with studies of homogeneous materials. For fast-neutron reactors, the materials of primary interest are iron, sodium, stainless steel, and graphite. These are the materials which have been used in a wide range of benchmark experiments with the apparatus described in Table XXXVII (Refs. 114 and 181–191). In addition, considerable attention has been paid to studies of heterogeneous layouts of these materials,[118,119,191-194] which are especially important in integral benchmark experiments for correcting cross sections in the design of shielding.

Radiation detectors

Activation detectors are used effectively in most laboratories as the basic detectors for measuring integral quantities. In measurements of energy dis-

Table XXXVII. Experimental apparatus used in benchmark studies of neutron shielding.

Number	Name of apparatus and site	Characteristic of source		Characteristic of shielding	Detectors	Comments
		Geometry and strength	Energy spectrum			
1	HARMONIE (Cadarache, France)	Cylindrical, 3 kW	Leakage spectrum from stainless-steel reflector	Fe and Na blocks; Fe + Na configuration	Activation detectors, recoil-proton counters	
2	TAPIRO (Casacia, Italy)	Cylindrical, 5–10 kW	Leakage spectrum from copper reflector	Na block, $1 \times 1 \times 1$ m	Activation detectors	
3	YAYOI (Tokyo University, Japan)	Cylindrical, 2 kW	Leakage spectrum from lead reflector	Fe blocks	Organic scintillator, proportional counters	Can be collimated into a broad or narrow beam
4	ASPIS (Winfrith, England)	Disk ($d = 1.1$ m), 7 W (2×10^{11} n/s)	Fission neutron spectrum	Fe block, $2 \times 2 \times 1.5$ m	Activation detectors, organic scintillator, proportional counters	Thermal-column converter
5	TRIGA converter (Casacia, Italy)	Disk, several watts	Fission neutron spectrum	Fe block, $1 \times 1 \times 1$ m	Activation detectors, fission chambers	
6	EURACOS II (Pavia, Italy)	Disk, 300 W	Fission neutron spectrum	Fe block, $1.5 \times 1.5 \times 1.5$ m	Activation detectors, fission chambers	
7	KND apparatus at the BR-10 reactor (Obninsk, USSR)	Disk ($d = 0.25$ m), 2 W	Fission neutron spectrum		Activation detectors	Thermal-column converter
8	^{252}Cf (Obninsk, USSR)	Point, $\sim 10^9$ n/s	Point, $\sim 10^9$ n/s	Fe sphere up to 0.7 m diameter; Na sphere up to 1 m in diameter	Single-crystal scintilllation spectrometer, multisphere spectrometer	

Table XXXVII. (continued)

Number	Name of apparatus and site	Characteristic of source		Characteristic of shielding	Detectors	Comments
		Geometry and strength	Energy spectrum			
9	^{252}Cf (Karlsruhe, FRG)	Point, 7×10^7 n/s	Fission neutron spectrum	Fe sphere up to 0.4 m in diameter	Recoil-proton counters, semiconductor spectrometer	Clean source geometry
10	^{252}Cf (Jaeri, Japan)	Point	Fission neutron spectrum	Fe sphere 0.5 m in diameter	Organic scintillator	Clean source geometry
11	14-MeV accelerator; $T(d,n)$ He reaction (Karlsruhe, FRG)	Point, 10^9 n/s	14 MeV	Fe cylinder		Clean source geometry
12	14-MeV accelerator; $T(d,n)$He reactor (Jaeri, Japan)	Point	14 MeV	Fe sphere 0.5 m in diameter	Time-of-flight spectrometer, NE-213 liquid scintillator	Clean source geometry
13	TSF (Oak Ridge, TN)	Spherical, 10 kW (broad beam in a concrete collimator)	Fission neutron spectrum softened by a moderator	Fe and steel blocks, sodium cylinder, 4.5 m, steel-sodium configuration	NE-213 scintillator; recoil-proton counters	Spherical geometry, can be collimated into a broad beam
14	B-2 apparatus at the BR-10 reactor (Obninsk, USSR)	Unidirectional disk ($d = 0.3$ m), $\sim 10^{13}$ n/s	Leakage spectrum from nickel reflector of a fast-neutron reactor	Steel-graphite, $1.25 \times 1.25 \times 0.88$ m	Activation detectors	
15	OR-M (Kurchatov, Institute, Moscow, USSR)	Disk, unidirectional ($d \leqslant 1$ m), 1.3×10^8(n/cm^2 s)	Leakage spectrum from graphite reflector of water-cooled, water-moderated reactor	Fe, Al, Be, etc., blocks, $1.2 \times 1.2 \times 0.5$ m	Organic scintillator	Can be collimated into a broad or narrow beam

tributions, activation detectors are widely supplemented with organic scintillators and proportional counters.[6] A set of activation reactions and corresponding cross sections was recommended at the meeting in Ref. 6 for use in benchmark experiments on shielding. This set includes the following reactions: $^{26}Al(n,\alpha)^{24}Na$; $^{32}S(n,p)^{32}P$; $^{58}Ni(n,p)^{58}Co$; $^{115}In(n,n')^{115m}In^{103}$;$Rh(n,n')$ ^{103m}Rh; $^{23}Na(n,y)^{24}Na$; $^{63}Cu(n,\gamma)^{64}Cu$; $^{55}Mn(n,\gamma)^{56}Mn$; $^{186}W(n,\gamma)^{187}W$; ^{197}Au $(n,\gamma)^{198}Au$; and $^{115}In(n,\gamma)^{116}In$.

Interpretation of benchmark experiments

The BNAB standard library of cross sections with the ROZ programs for calculating radiation transport is widely used in the Soviet Union for the interpretation of benchmark experiments. The ZAKAT + ROZ-11 programs is used for sensitivity calculations (Sec. 2.2).

For a comparison of data from benchmark experiments with calculated results, many non-Soviet laboratories use the EURLIB standard 100-group cross-section library for calculations. That library was generated from the EMDF/BIII and ENDF/BIV files by the SUPERTOG program for generating group cross sections. The ANISN and DOT programs, which implement the discrete S_N method, and the MORSE program, with the Monte Carlo method, are widely used as standard programs for calculations on radiation transport. The SWANLAKE + ANISN combination is used for sensitivity calculations.

Benchmark calculation models

In addition to the program of benchmark experiments, two benchmark calculation models have been proposed in foreign countries for testing various calculation methods and for determining the effect of using various "homebrew" cross-section libraries in calculations of relative sensitivities. In addition, the comprehensive study of benchmark calculation models has made it possible for various laboratories to adopt standard software for sensitivity calculations, the SWANLAKE + ANISN package.

The models proposed at the research centers at Cadarache and Stuttgart describe, respectively, a one-dimensional steel-sodium radial shield for a fast reactor[195] and a two-dimensional radial shield for a water-cooled, water-moderated power reactor, along with its one-dimensional representation.[196] The benchmark calculation models are described in detail in Chap. 2, Sec. 3 (see Table V, models 10 and 11). The results of a study of these benchmark models are reproduced in part in Chap. 3, Sec. 4.

In the Soviet Union, similar work on benchmark models was discussed at the Second All-Union Scientific Conference on Shielding against Ionizing Radiations in Nuclear-Engineering Installations. Before the Conference, four test problems were sent to various laboratories, along with corresponding specifications regarding shielding calculations. After the calculations were carried out in the various laboratories, the results were compared with the data of the corresponding benchmark experiments.[2]

Table XXXVIII. Energy distribution of the source neutrons for the 26-group partition.

Group	Group boundaries, MeV	ΔU	φ_i (n/cm² s) in the group
1	6.5–10.5	0.48	3.49 + 7[a]
2	4.0– 6.5	0.48	1.28 + 8
3	2.5– 4.0	0.48	3.20 + 8
4	1.4– 2.5	0.57	1.35 + 9
5	8.0 − 1– 1.4	0.57	2.69 + 9
6	4.0 − 1– 8.0 − 1	0.69	5.14 + 9
7	2.0 − 1– 4.0 − 1	0.69	4.17 + 9
8	1.0 − 1– 2.0 − 1	0.69	1.63 + 9
9	4.65 − 2– 1.0 − 1	0.77	1.02 + 9
10	2.15 − 2– 4.65 − 2	0.77	8.02 + 8
11	1.00 − 2– 2.15 − 2	0.77	8.63 + 8
12	4.65 − 3– 1.00 − 2	0.77	8.83 + 8
13	2.15 − 3– 4.65 − 3	0.77	8.83 + 8
14	1.00 − 3– 2.15 − 3	0.77	8.83 + 8
15	4.65 − 4– 1.00 − 3	0.77	8.83 + 8
16	2.15 − 4– 4.65 − 4	0.77	8.50 + 8
17	1.00 − 4– 2.15 − 4	0.77	7.53 + 8
18	4.65 − 5– 1.00 − 4	0.77	7.04 + 8
19	2.15 − 5– 4.65 − 5	0.77	6.40 + 8
20	1.00 − 5– 2.15 − 5	0.77	5.75 + 8
21	4.65 − 6– 1.00 − 5	0.77	5.11 + 8
22	2.15 − 6– 4.65 − 6	0.77	4.17 + 8
23	1.00 − 6– 2.15 − 6	0.77	3.52 + 8
24	4.65 − 7– 1.00 − 6	0.77	2.56 + 8
25	2.15 − 7– 4.65 − 7	0.77	6.60 + 7
26	0.252 −07		3.18 + 7

[a] Read $a \pm b$ as $a \times 10^{\pm b}$.

These benchmark experiments consisted of an experiment on the transport of high-energy γ radiation through various shielding materials, carried out in a special radiation loop with ^{16}N of the LIDO reactor at Harwell,[197] and three experiments on measurements of neutron spectra.

Let us examine these benchmark test experiments on measurements of neutron spectra.[2] In a first experiment, a thin converter of natural uranium served as a uniform, isotropic, disk-shaped ($R = 59.3$ cm) source of neutrons with a fission spectrum. The total number of fission events per unit time in the converter was 2.27×10^7 s^{-1}. The neutron yield per fission event was assumed to be 2.43. A shield with transverse dimensions of 191×183 cm consisted of 24 steel plates each 5.08 cm thick. The atomic densities of the components of the steel were as follows, in units of 10^{24} atoms/cm^3: H, 2.071×10^{-5}; C, 7.996×10^{-4}; Mn, 5.660×10^{-4}; Fe, 7.404×10^{-2}. A benchmark experiment in the geometry described above was carried out in an apparatus for studying the ASPIS shield at the NESTOR reactor in Winfrith.[187] The spatial-energy flux densities of neutrons were measured in steel shielding at thicknesses of 20.3, 50.8, 76.2, and 101.6 cm. The detectors were a hydrogen counter, a scintillation spectrometer with NE-213 liquid scintillator, and activation detectors.

In a second experiment, a collimated neutron beam from the core of a reactor produced a unidirectional disk-shaped neutron source 30 cm in diameter at the entrance to a heterogeneous shielding prism. The energy distribution of the neutrons of this source, shown in a group representation in Table XXXVIII, was similar to a fission spectrum at energies above 3 MeV, while significantly softer than a fission spectrum at lower energies. This shield consisted of five alternating layers of 1Kh18N9T stainless steel, each 16 cm thick, and graphite, each 20 cm thick. The transverse dimensions of the shield were 125×125 cm, and its total thickness was 88 cm. The atomic densities of the components of the stainless steel were, in units of 10^{24} atoms/cm^3: C, 9.62×10^{-5}; Si, 4.15×10^{-4}; Ti, 4.52×10^{-4}; Ni, 8.21×10^{-3}; Cr, 1.47×10^{-2}; and Fe, 5.86×10^{-2}. The atomic density of the graphite was 8.03×10^{22} atoms/cm^3. Measurements of the energy distributions of the neutron flux density along two mutually perpendicular axes made it possible to convert the experimental data to a one-dimensional plane geometry with a unidirectional plane source emitting neutrons normal to the surface of the heterogeneous shield. This benchmark experiment was carried out in the B-2 beam of the BR-10 fast-neutron reactor.[119,194] The neutron spectra were measured with a set of 12 threshold detectors and seven resonance activation detectors at shielding thicknesses of 6.5, 25, 42.5, and 61 cm.

In a third experiment,[198] a unidirectional monoenergetic beam of deuterons with an energy of about 400 keV passed through a channel in a spherical shield and struck a tritium target at the center of a cavity in a spherical iron shield (the cavity radius was $a_0 = 0.54$ cm). Neutrons with an energy of 15 MeV were emitted as a result of the $T(d,n)^4$He reaction. A time-of-flight method was used to study the energy distribution of the neutrons escaping from spheres with radii $r = 4.46, 13.46$, and 22.40 cm along the line connecting the source and the detector ($\theta = 0°$). It was assumed that an isotropic point source was at the center of a cavity with a radius $a_0 = 0.54$ cm and was emitting one neutron per second. The composition of the spheres of iron ($\rho = 7.86$ g/cm^3) was as follows (percent by mass): Fe, 98.50; Mn, 0.50; C, 0.50; Si, 0.37; P, 0.04; S, 0.04; and other elements, 0.05. The neutron detectors were a scintillation spectrometer with NE-213 liquid scintillator for detecting neutrons with energies above 2 MeV and a ^{6}Li detector for detecting neutrons with energies from 1 MeV down to several kiloelectron volts.

2. Comparison of calculations and experiments; roles played by various components of the calculation errors

The results of the calculations carried out in the various laboratories were compared in Ref. 2 with the data of the three benchmark experiments described in Sec. 4.1. Some general preliminary conclusions were drawn: (1) In all the cases considered, the discrepancies between the calculated and experimental results exceeded the errors estimated in the calculated results by the authors of the calculations. (2) The discrepancies between the calculated and experimental results averaged 30%–70% for integral values and a factor of about 3 to 5 for the differential characteristics of the radiation field.

These comparisons amount to only a first step in the testing of the calculation methods and constants which were being used. Continuous further work is required to develop benchmark problems and to make comparisons.

To illustrate the purposes of the benchmark experiments, we consider the example of the interpretation of benchmark experiments for iron and a heterogeneous steel-graphite shield to test the ZAKAT + ROZ-11 and SWANLAKE + ANISN sensitivity-analysis programs. We also consider the example of the use of experimental and calculated results with a set of benchmark integral experiments to correct cross sections during the design of shielding.

The ZAKAT + ROZ-11 and SWANLAKE + ANISN program packages were used for an interpretation of a benchmark experiment in spherical geometry for an iron shield with a californium source at its center. In the experiments,[114] six different iron spheres, with diameters of 15, 20, 25, 30, 35, and 40 cm, were used. High-purity iron was used (the impurity concentrations were 0.07% C, 0.05% Mn, 0.009% P, and 0.007% S). The ^{252}Cf source, with an activity of 7×10^7 n/s, was in a capsule at the center of the iron sphere. The detectors were positioned 1 m from the center of the sphere on an axis in the radial direction. The angular spectra of leakage neutrons from the surface of the sphere were measured over the energy range from 60 keV to 5 MeV.

The spectra of leakage neutrons from the surface of the iron sphere 40 cm in diameter were calculated for this geometry of the benchmark experiment. In addition, the relative sensitivities of the calculated results to the neutron interaction cross sections were calculated for iron spheres with radii of 20 cm (as in the benchmark experiment) and 100 cm.

The ANISN calculations were carried out in the $S_{12}P_3$ approximation with the help of the standard 100-group EURLIB system.

The ROZ-11 calculations were carried out in the P_5P_5 approximation on a BÉSM-6 computer with the help of the ARAMAKO-2F 26-group system.

Figure 53 shows the measured and calculated energy distributions of the neutron current density at the surface of an iron sphere 20 cm in radius. We see that the experimental results agree with the calculated results found by the ROZ-11 and ANISN programs with the corresponding systems of group cross sections.

Figures 54(a) and 54(b), compare the energy dependence of the relative sensitivity of the neutron current density to the total iron cross section for a sphere 20 cm in radius and the neutron flux density for a sphere 100 cm in radius according to calculations by the ZAKAT + ROZ-11 and SWANLAKE + ANISN software. From Fig. 54(a) we see an agreement in the energy dependence of the sensitivity in terms of the shape of the curves, while there is some discrepancy (by up to a factor of 2) in absolute values. The relative sensitivity (the sum over all groups) is 0.05 for the ZAKAT + ROZ-11 software and 0.13 for the SWANLAKE + ANISN software. This discrepancy in terms of the integral sensitivity leads, under the assumption that the sensitivity is a linear function of the partial cross sections, to an 8% change in the functional in the

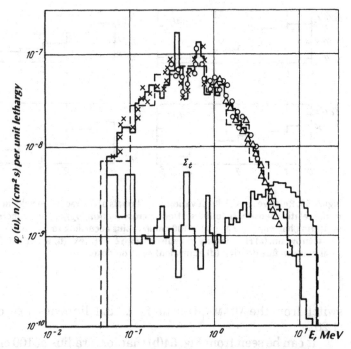

Figure 53. Energy distributions of the neutron current density at the surface of an iron sphere 20 cm in radius. $\times, \bigcirc, \triangle$: experimental data; —— and -----: results calculated by the ANISN and ROZ-11 programs, respectively. Shown for comparison is the energy dependence of the total cross section Σ_t.

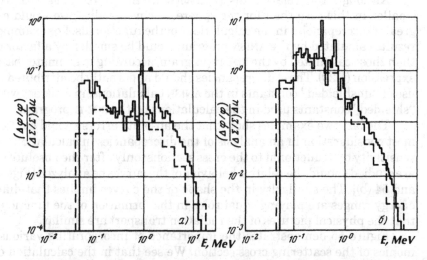

Figure 54. Energy dependence of the relative sensitivity of (a) the current density of neutrons with $E > 60$ keV for a sphere 20 cm in radius and (b) the flux density of neutrons with $E > 60$ keV for a sphere 100 cm in radius to the total iron cross section, according to calculations by the ZAKAT program (dashed line) and the SWANLAKE program (solid line).

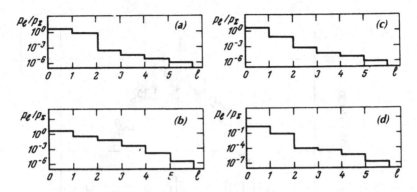

Figure 55. Relative sensitivity of various functionals of the radiation field, normalized to the relative sensitivity to the scattering cross section, p_e/p_Σ, as a function of the scattering harmonic for a sphere 100 cm in radius according to calculations by the ROZ-11 program. (a) Flux density of neutrons with $E > 50$ keV; (b) with $E > 1.4$ MeV; (c) total neutron flux density; (d) equivalent neutron dose.

switch from the ARAMAKO-2F to the EURLIB libraries according to expression (1.20).

It can be seen from Fig. 54(b) that for a radius of 100 cm the discrepancy between the results does not exceed the errors in the calculated data; i.e., the energy dependences of the relative sensitivity obtained by the different methods agree better in terms of the shape of the curves and in terms of absolute values. The relative sensitivity is -5.42 for the ZAKAT + ROZ-11 software and -6.27 for the SWANLAKE + ANISN software.

Although the relative discrepancies in the integral sensitivity are smaller in this case than for the sphere 20 cm in radius, we would expect greater discrepancies in the calculations of the functionals. For example, the results calculated by the ANISN program would be smaller by a factor of 2.3 than those calculated by the ROZ-11 program, according to estimates based on expression (1.20). These discrepancies should apparently be attributed to the use of "unshielded" constants in the ANISN calculations, in contrast with the "shielded" constants used in the calculations by the ROZ-11 program.

The last two examples show that the absolute errors should be the foremost consideration in an analysis of the discrepancies in calculations of the sensitivity of a functional to the cross sections; only after the absolute errors are studied should the relative behavior of the curves be analyzed [Figs. 54 (a) and 54 (b)]. The similarity in the shape of the curves implies that different energy ranges are playing equal roles in the formation of the functional, so that the physical pictures of the radiation transport are similar.

Figure 55 demonstrates the importance of incorporating various harmonics of the scattering cross section. We see that in the calculation of the total flux density and of the equivalent neutron dose it is sufficient in practice to consider only the zeroth and first harmonics of the scattering characteristic. In calculations of the flux density of fast neutrons, the second and third harmonics must also be taken into consideration.

Figure 56 shows relative sensitivities of various functionals of the neu-

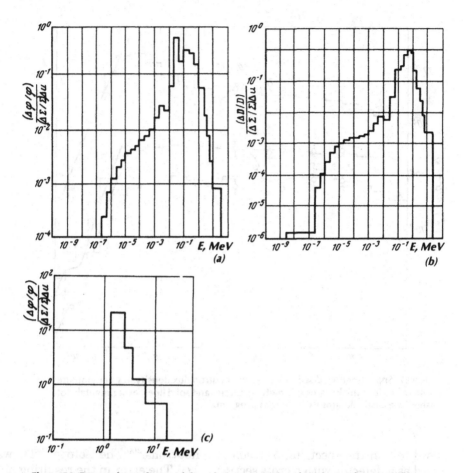

Figure 56. Energy dependence of the relative sensitivity of (a) the total neutron flux, (b) the equivalent dose, and (c) the fast-neutron flux to the total Fe cross section for a sphere 100 cm in radius.

tron field to the total cross section of iron according to ZAKAT calculations. The greatest sensitivity for the total flux density and for the equivalent neutron dose is observed at neutron energies in the range 20–800 keV. Consequently, for thick layers of iron (as expected) the process which is basically responsible for forming these functionals is the elastic scattering of intermediate-energy neutrons.

The ZAKAT + ROZ-11 software was also used for calculations on the benchmark experiment (described in Sec. 4.1) on neutron propagation in a heterogeneous steel-graphite shield, carried out in the B-2 beam of the BR-10 reactor.[119,194]

The experiments used a set of 12 threshold detectors and seven resonance activation detectors in cadmium or boron shields. The activity was measured by a scintillation γ spectrometer with a NaI(Tl) crystal and with a 4π flowing-gas β counter. The spectra of fast neutrons were reconstructed by the "rapid" method of Refs. 199 and 200 and by a modified maximum-likelihood method[52] (MMP-M). The method of subtraction of the $1/v$ part of the cross section was

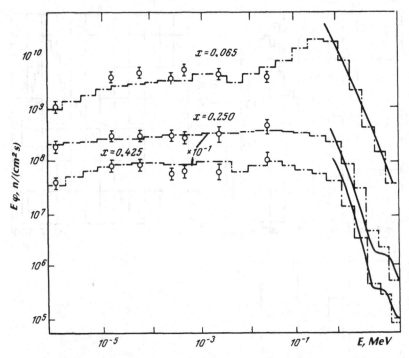

Figure 57. Spatial-energy distributions of the neutron flux density in steel-graphite shields of various thicknesses x (in meters). Circles and solid lines, experimental; dot-dashed lines, calculations by the ROZ-11 program.

used to find the spectrum of resonance neutrons.[200] The isotope ^{164}Dy was used as a detector with a cross section $\sim 1/v$. The error in the resulting neutron energy spectra was due primarily to two factors: errors in the determination of the activity per nucleus at saturation and errors in the method used for unfolding the spectra.

Among the basic components of the errors in the determination of the activity per nucleus at saturation are (a) the statistical error in the determination of the count rate of β particles and γ rays (this error ranges from 1% to 10%), (b) errors in the measurement of the mass of the tracer ($< 1\%$), (c) errors in the measurements of the irradiation time and the reactor power ($\sim 5\%$, reduced by monitoring the reactor power during the irradiation), (d) errors in the readings of the detectors ($< 3\%$), and (e) errors in the approximate calculation of the integral in the transformation from a disk-shaped unidirectional source to a plane source (5%–10%).

In an effort to reduce the resultant error in the determination of the activity at saturation, measurements were repeated many times; as a result, the error in the measurement of the activity at saturation was 5%–10%. An uncertainty of 10% leads to an error of 20% to 30% in the unfolded spectra of the fast neutrons when the rapid method of Ref. 200 is used or an error of no more than 30% when the MMP-M maximum-likelihood method is used.[52] For resonance neutrons, the maximum error in the unfolding of the spectrum by

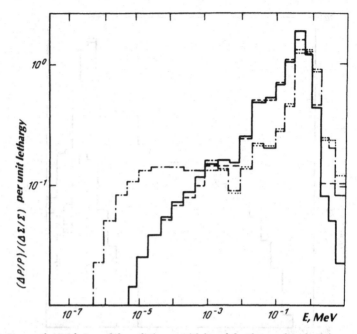

Figure 58. Energy dependence of the relative sensitivity of the tissue-absorbed dose rate behind a steel-graphite shield to the cross sections for the interactions of neutrons with carbon (solid line, ZAKAT calculations; dashed line, direct-substitution method) and with steel (dot-dashed line, ZAKAT calculations; dotted line, direct-substitution method).

the method involving subtraction of the $1/v$ part of the cross section is no greater than 20% at a 5% error in the activity measurement.[200]

In summary, neutron-energy spectra over the range from 1 eV to 10 MeV were obtained in the experiment with an error no greater than $\sim 30\%$. The standard geometry and the measurement procedure make it possible to classify this experiment as a benchmark experiment and to use it as a test of calculation programs and of cross sections of the interaction of radiation with matter.

Figure 57 shows spatial-energy distributions of the neutron flux density found in the benchmark experiment and in calculations by the ROZ-11 program with the ARAMAKO-2F system of cross sections. The experimental and calculated results agree within the experimental errors of up to 30%.

The software mentioned above was used to calculate the relative sensitivities of the absorbed dose rate in tissue P to the interaction cross sections of all the components of the shielding and to the source and detector specification functions.

Figure 58 shows the energy dependence of the relative sensitivity of P to the total cross sections of steel and graphite found by the ZAKAT program and by the direct-substitution method with a 5% variation in all the partial cross sections. Additional calculations by the direct-substitution method, for various changes in the partial cross sections, revealed that linear perturbation theory is valid (within $\sim 10\%$) for variations $\pm 5\%$ in the cross sections. These results show that 75% of the dose comes from neutrons with energies

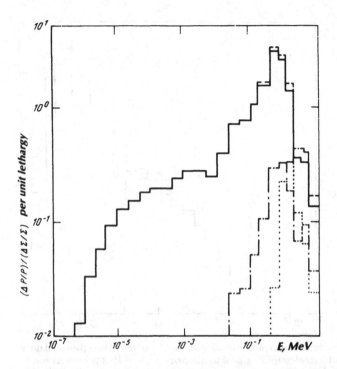

Figure 59. Energy dependence of the relative sensitivity of the absorbed dose rate to the cross section for elastic scattering (solid line), to the zeroth harmonic of the elastic cross section (dashed line), to the first harmonic of the elastic cross section (dot-dashed line), and to the inelastic cross section (dotted line).

2.15×10^{-2}–2.5 MeV, to a large extent because the neutrons in this energy range are predominant in the source spectrum and because of the particular shape of the detector response function. Here 55% of the dose is determined by interactions of neutrons with the graphite, and 45% by interactions with the steel.

Figure 59 shows the energy dependence of the relative sensitivity of P to the elastic and inelastic cross sections and to harmonics of the scattering cross section. It follows from these results that inelastic scattering makes a minor contribution ($\sim 0.6\%$) to the dose; the particular shape of the source spectrum is quite influential here. A decisive role is played by scattering processes with a small anisotropy: The relative sensitivity of the dose to the zeroth harmonic is -9.94; that to the first harmonic is $+0.73$; and that to the sum of the second through fifth harmonics is $+0.07$. It follows that ignoring harmonics from the second through the fifth will result in only a 7% underestimate of the dose.

Figure 60 shows the energy dependence of the relative sensitivity of the absorbed dose rate in tissue to the source specification function and the detector response function in the group representation. We see that 94% of the dose for this particular shield layout is determined by source neutrons with energies in the range 0.2–4.0 MeV, while for the neutrons detected this range

Figure 60. Energy dependence of the relative sensitivity of the absorbed dose rate to the source specification function (solid line) and to the detector response function (dashed line).

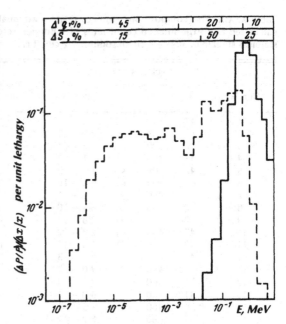

is considerably wider (from 10^{-6} to 1.4 MeV). The irregularities in the energy dependence of the relative sensitivity of the dose to the detector response function for the interval 10^{-3}–10^{-1} MeV stem from the minima and maxima in the cross sections for the interaction of neutrons with the components of the steel.

Let us use the results found on the relative sensitivities to estimate the errors caused in the calculations of the dose rate by errors in the source specification function, the detector response function, and the neutron interaction cross sections. Figure 60 shows the calculated relative errors (ΔS) in the source specification function and data on the relative errors (Δq) in the detector response function. The errors in the calculation of P which stem from ΔS and Δq are then 7% and 11%, respectively. The calculation error which results from the errors in all the partial interaction cross sections is about 25%. It follows from these estimates of the dose error that the largest errors stem from uncertainties in the neutron interaction cross sections, as has been pointed out previously.

Several integral benchmark experiments have been carried out over the past decade in the U.S.A., and many of these experiments serve as criteria for evaluating the cross sections of the ENDF/B library.[183,184] However, except for the last series of experiments, with iron-sodium-steel shields, carried out at the TSF installation, the results of these experiments have not been analyzed comprehensively, and they have not been used to adjust the constants. The reason is that the errors in this integral experimental information have not been carefully evaluated, and possible correlations between different pieces of experimental information have not been estimated.

Consequently, a complete study was carried out[118,201] for a series of inte-

Table XXXIX. The compositions studied, relative attenuation factors, and comparison of the calculated and experimental count rates of a Bonner detector. The shields consist of x cm of stainless steel (No. 304) + y cm of sodium + z cm of carbon steel.

	Layer thickness, cm						
Number	x	y	z	Attenuation[a]	Measurement error, %	$c(51)/e$	$c(171)/e$[b]
1	47	155	0	1.2 — 3[c]	5	0.90	1.05
2	47	309	0	2.0 — 5	5	0.78	0.84
3	47	309	15	2.6 — 7	5	1.04	0.94
4	47	309	31	6.1 — 9	5	1.13	1.29
5	47	309	46	1.4 — 9	5	0.87	1.52
6	47	309	62	6.4 — 10	8	0.75	1.37
7	47	460	0	1.6 — 7	5	0.84	0.73
8	47	460	15	8.5 — 10	8	1.54	0.95
9	47	460	31	7.8 — 12	10	1.82	1.38
10	47	460	41	2.8 — 12	10	0.46	0.80
11	47	460	51	1.5 — 12	10	0.30	0.72
12	47	460	62	8.2 — 13	15	0.30	0.75
13	31	460	0	4.0 — 7	5	0.88	0.80
14	31	460	15	2.4 — 9	5	1.51	1.07
15	31	460	31	4.0 — 11	8	1.38	1.42
16	31	460	41	1.5 — 11	8	0.82	1.24
17	31	460	51	8.2 — 12	10	0.77	1.26
18	31	460	62	4.8 — 12	10	0.76	1.26
19	31	460	72	2.8 — 12	10	0.77	1.30
20	31	460	82	1.6 — 12	10	0.73	1.25
21	31	460	90	1.1 — 12	15	0.70	1.20

[a] Defined as the ratio of the measured detector response behind the configuration to the calculated response at the point corresponding to the beginning of the stainless steel plate.
[b] Normalized ANISN calculations. The normalization error is estimated to be ± 10%, so that the possible error in the quantity $c(171)/e$ is ± 15%.
[c] Read 1.2×10^{-3}.

gral benchmark experiments to model the shielding of a fast-neutron reactor. The experimental layouts were chosen for the design of the upper axial shield of the Clinch River fast-neutron reactor. The primary motivation for these studies was to determine the validity of the nuclear data files for Na and Fe in the ENDF/B-IV library.

Experiments carried out at the TSF installation at the Oak Ridge National Laboratory (Sec. 4.1) dealt with three basic shield layouts: (1) 47 cm of stainless steel + 309 cm of Na + up to 62 cm of Fe; (2) 47 cm of stainless steel + 460 cm of Na + up to 62 cm of Fe; (3) 31 cm of stainless steel + 460 cm of Na + up to 90 cm of Fe.

In each layout, the components were installed in turn, and measurements were taken behind the installed part of the layout before the following components were installed. For the measurements, a detector was placed in an air-filled gap just beyond the partially installed shield. A lithium hydride plate 46 cm was placed behind this detector; this plate, along with additional peripheral shielding, reduced the background from particles which have penetrated the shield and from scattering by the earth and the air.

The detector used for these measurements was a 10-cm Bonner detector

Table XL. Results of a correction of the basic cross sections, %.

Reaction	< 1 keV	1–250 keV	~300 keV	1–5 MeV	5–15 MeV
Na(n,n)	− 8	+ 6	− 13	+ 4	+ 5
Na(n,n')	—	—	—	− 3	− 3
Fe(n,n)	+ 5	—	—	− 5	+ 6
Fe(n,n')	—	—	—	+ 8	− 7

consisting of a spherical BF_3 counter with a diameter of 5.08 cm inside a polyethylene shell 2.54 cm thick and a thin cadmium shell. This detector was most sensitive to neutrons in the energy range from 1 eV to 100 keV and was convenient for detecting the low-energy neutrons which leaked from the mockups of the upper axial shield.

Calculations were carried out for all the experimental shields on the basis of the DOT-III two-dimensional program,[202] based on the discrete-ordinates method, and a 51-group library of cross sections. This library was generated from the ENDF/B-IV data for the calculations on the design of a shield for a fast-neutron reactor. This analysis also used a standard shielding library with detailed group representations of the cross sections in 171 groups for one-dimensional calculations by the ANISN program. The results found by the ANISN program were used to supplement the DOT calculations through the incorporation of effects of the structural details of the cross sections, which are not preserved correctly in the 51-group library. In addition, a one-dimensional sensitivity analysis was carried out for all the layouts with a 171-group library of cross sections.

Table XXXIX compares the experimental data found (the measurement errors are also given) with results calculated by the DOT program [the column headed c(51)/e] and the ANISN program [c(171)/e].

For the calculations with 51 groups, the agreement with experiment is within ± 30% nearly everywhere except for points behind shields containing 47 cm of stainless steel + 460 cm of Na + z cm of Fe, where there is an overestimate by a factor of 1.8 for $z = 31$ cm, and there is an underestimate by a factor of 3.3 for $z \geqslant 51$ cm.

The results of the 51-group DOT-III calculations were corrected on the basis of the 171-group ANISN calculations, and the results were used to estimate the ratio c/e (calculated/experimental) for the library of cross sections with the detailed group structure. Analysis of the corrected values of c/e reveals an overall improvement in the agreement of the data in comparison with the 51-group results. Most of the values of c/e now agree within an error of 30% everywhere except for the points beyond 30 cm of Fe, where the agreement is only slightly poorer. A significant improvement in the results behind the shield consisting of 47 cm on stainless steel + 460 cm of Na + 62 cm of Fe is observed primarily because of the incorporation of several groups near the minimum of the total Na cross section at 300 keV.

The FORSS program was used to analyze the errors for all 21 layouts.[57,203] The files of errors in the cross sections of the ENDF/B-IV library for Fe and Na which were evaluated at the Oak Ridge National Laboratory were combined with information on the sensitivity to calculate the errors in the calculated readings of a Bonner detector due to the errors in the data on the cross sections. The results showed that for all the experimental studies there are standard deviations ($1\,\sigma$) $\sim 50\%$. As a result of an analysis of the errors, the basic cross sections of Na and Fe were corrected; the results are shown in Table XL (Ref. 201). The most general conclusion which can be drawn from this study is that the use of the integral data found from the measurements at the TSF in procedures for correcting cross sections significantly reduces the calculation errors in work on the design of shielding for fast-neutron reactors.

References

1. *Metrologiya neĭtronnogo izlucheniya na reaktorakh i uskoritelyakh. Trudy II Vsesoyuznogo soveshchaniya (Metrology of the Neutron Radiation at Reactors and Accelerators*, Proceedings of the Second All-Union Conference), edited by R. D. Vasil'ev (VNIIFTRI, Moscow, 1974), Vol. 2.

2. V. V. Bolyatko, M. Yu. Vyrskiĭ, V. P. Mashkovich, and V. K. Sakharov, in *Voprosy dozimetrii i zashity ot izlucheniĭ (Questions of Radiation and Dosimetry Shielding), No. 18*, edited by V. K. Sakharov (Atomizdat, Moscow, 1979), pp. 6–19.

3. *Proceedings of the Specialist's Meeting on Sensitivity Studies and Shielding Benchmarks, Paris, 7–10 October 1975* (IAEA, Vienna, 1975).

4. *Proceedings of the Specialist's Meeting on Differential and Integral Nuclear Data Requirements for Shielding Calculations, Vienna, 12–15 October 1976* (IAEA, Vienna, 1978).

5. Materials used in the RSIC Workshop on the ORNL FORSS Sensitivity and Uncertainty Analysis Code System, 23–24 August 1978, Oak Ridge, TN.

6. J. Butler and R. Nicks, in *Sensitivity Studies and Shielding Benchmark Experiment*, Report of a Joint NEA/EURATOM Specialist Meeting Held at Ispra, April 1974, NEACRP Report No. L-I.

7. Proceedings of the Fourth International Conference on Reactor Shielding, Paris, 9–13 October 1972 (unpublished).

8. *Proceedings of the Fifth International Conference on Reactor Shielding*, RSIC ORNL, edited by R. W. Roussin, L. S. Abbott, and D. E. Bartine (Science Press, Princeton, NJ, 1977).

9. *Sbornik tezisov dokladov vsesoyuznoĭ nauchnoĭ konferentsii po zashite ot ioniziruyushikh izlucheniĭ yaderno-tekhnicheskikh ustanovok (Proceedings of the All-Union Scientific Conference on Ionizing-Radiation Shielding in Nuclear-Engineering Installations), 17–19 December 1974* (MIFI, Moscow, 1974).

10. *Sbornik tezisov dokladov Vtoroĭ Vsesoyuznoĭ nauchnoĭ konferentsii po zashite ot ioniziruyushikh izlucheniĭ yaderno-tekhnicheskikh ustanovok (Proceedings of the Second All-Union Scientific Conference on Ionizing-Radiation Shielding in Nuclear-Engineering Installations), 19–21 December 1978* (MIFI, Moscow, 1978).

11. *Sbornik tezisov dokladov Tret'eĭ vsesoyuznoĭ nauchnoĭ konferentsii po zashite ot ioniziruyushikh izlucheniĭ yaderno-tekhnicheskikh ustanovok (Proceedings of the Third All-Union Scientific Conference on Ionizing-Radiation Shielding in Nuclear-Technology Installations), 27–29 October 1981* (TGU, Tbilisi, 1981).

12. E. A. Straker, *Sensitivity of Neutron Transport in Oxygen to Various Cross-Section Sets*, ORNL Report No. ORNL-TM-2252, 1968.

13. D. E. Bartine, E. M. Oblow, and F. R. Mynatt, *Neutron Cross-Section Sensitivity Analysis. A General Approach Illustrated for a Na-Fe System*, ORNL Report No. ORNL-TM-3944, 1972.

14. H. Goldstein *et al.*, in Proceedings of the Fourth International Conference on Reactor Shielding, Paris, 9–13 October 1972 (unpublished), p. 1.

15. D. E. Bartine, F. R. Mynatt, and E. M. Oblow, SWANLAKE: *A Computer Code Utilizing ANISN Transport Calculations for Cross-Section Sensitivity Analysis*, ORNL Report No. ORNL-TM-3809, 1973.

16. E. M. Oblow, *General Sensitivity Theory for Radiation Transport*, ORNL Report No. ORNL-TM-4110, 1973.

17. D. E. Bartine, E. M. Oblow, and F. R. Mynatt, Nucl. Sci. Eng. **55**, 147 (1974).

18. H. Goldstein, L. Y. Huang, L. P. Ky, and D. Cacuci, Transl. Am. Nucl. Soc. **18** (1), 364 (1974).

19. D. V. Markovskiĭ, G. E. Shatalov, and G. B. Yan'kov, Preprint No. IAE-2579, IAE im. I. V. Kurchatova, Moscow, 1975.

20. T. A. Germogenova, A. P. Suvorov, and V. A. Utkin, in *Proceedings of the Specialist's Meeting on Differential and Integral Nuclear Data Requirements for Shielding Calculations, Vienna, 12–15 October 1976* (IAEA, Vienna, 1978), pp. 563–572.

21. Yu. A. Medvedev, E. V. Metelkin, B. M. Stepanov, and G. Ya. Trukhanov, in *Yaderno-fizicheskie konstanty v prikladnykh zadachakh neĭtronnoĭ fiziki (Nuclear-Physics Constants in Applications of Neutron Physics)* (VNIIOFI, Moscow, 1976), pp. 32–60.

22. E. M. Oblow, *Survey of Shielding Sensitivity Analysis Development and Applications Program at ORNL*, ORNL Report No. ORNL-TM-5176, 1976.

23. M. L. Williams and W. W. Engle, Nucl. Sci. Eng. **62**, 92 (1977).

24. S. A. W. Gerstl, D. J. Dudziak, and D. W. Muri, Nucl. Sci. Eng. **62**, 137 (1977).

25. M. L. Williams, Nucl. Sci. Eng. **63**, 220 (1977).

26. M. L. Williams and W. W. Engle, in *Proceedings of the Fifth International Conference on Reactor Shielding*, RSIC ORNL, edited by R. W. Roussin, L. S. Abbott, and D. E. Bartine (Science Press, Princeton, NJ, 1977), pp. 770–776.

27. F. G. Perey, *The Data Covariance Files for ENDF/B-V*, Report No. ENDF-249, ORNL, 1977.

28. A. M. Zhezlov *et al.*, At. Energy **40**, 313 (1980).

29. G. N. Manturov, V. I. Savitskiĭ, and A. I. Ilyushkin, in *Neĭtronnaya fizika. Materialy V Vsesoyuznoĭ konferentsii po neĭtronnoĭ fizike (Neutron Physics. Proceedings of the Fifth All-Union Conference on Neutron Physics, Kiev, 15–19 September 1980)*, edited by L. N. Usachev (TsNIIAI, Moscow, 1981), Pt. 3, p. 323.

30. V. V. Bolyatko *et al.*, At. Energy **50**, 328 (1981).

31. L. N. Usachev, in *Materialy mezhdunarodnoĭ konferentsii po mirnomu ispol'zovaniyu atomnoĭ énergii (Proceedings of the International Conference on the Peaceful Use of Atomic Energy, Geneva, 1955)* (Gosénergoizdat, Moscow-Leningrad, 1958), Vol. 5, p. 598.

32. B. B. Kadomtsev, Dokl. Akad. Nauk SSSR **117** (3) 541 (1957).

33. V. V. Orlov, in *Voprosy fiziki yadernykh reaktorov. Trudy FEI (Questions of the Physics of Nuclear Reactors. Proceedings of the FEI)* (Obninsk, 1968), No. 1, p. 38.

34. A. A. Abagyan, V. V. Orlov, and G. N. Rodionov, in *Voprosy fiziki zashity reaktorov (Questions of the Physics of Reactor Shielding)*, edited by D. L. Broder *et al.* (Gosatomizdat, Moscow, 1963), pp. 7–24.

35. A. A. Abagyan *et al.*, in *Trudy Tret'eĭ mezhdunarodnoĭ konferentsii po ispol'zovaniyu atomnoĭ énergii v mirnykh tselyakh (Proceedings of the Third International Conference on the Use of Atomic Energy for Peaceful Purposes, Geneva, 1964)*, multilingual ed. (United Nations, New York, 1965), Vol. 4, p. 359.

36. L. N. Usachev, At. Energy **15**, 472 (1963).

37. L. N. Usachev and S. M. Zaritskiĭ, Byull. Inf. Tsentra. Yadernym Dennym, No. 2, 242 (1965).

38. E. A. Stumbur, *Primenenie teorii vozmusheniĭ v fizike yadernykh reaktorov (Applications of Perturbation Theory in the Physics of Nuclear Reactors)* (Atomizdat, Moscow, 1976).

39. L. N. Usachev and Yu. G. Bobkov, *Teoriya vozmusheniĭ i planirovanie éksperimenta v probleme yadernykh dannykh dlya reaktorov (Perturbation Theory and Experimental Planning in the Problem of Nuclear Data for Reactors)* (Atomizdat, Moscow, 1980).

40. A. M. Weinberg and E. P. Wigner, *Physical Theory of Neutron Chain Reactors* (Univ. of Chicago Press, Chicago, 1958), Russ. translation published by Izd-vo inostr. lit., Moscow, 1961.

41. E. P. Wigner, in *Theory of Nuclear Reactors*, Russian translation (Gosatomizdat, Moscow, 1963), p. 103.

42. R. Ehrlich and H. Hurwitz, Nucleonics **12** (2), 23 (1954).

43. S. Glasstone and M. Edlund, *Fundamental Theory of Nuclear Reactors*, Russian translation (Izd-vo inostr. lit, Moscow, 1954).

44. G. I. Marchuk and V. V. Orlov, in *Neĭtronnaya fizika (Neutron Physics)* (Gosatomizdat, Moscow, 1961), p. 30.

45. G. I. Marchuk and V. I. Lebedev, *Chislennye metody v teorii perenosa neĭtronov (Numerical Methods in Neutron Transport Theory)*, 2nd ed. (Atomizdat, Moscow, 1981).

46. V. V. Orlov *et al.*, in *Voprosy fiziki zashity reaktorov (Questions of the Physics of Reactor Shielding)*, edited by D. L. Broder *et al.* (Atomizdat, Moscow, 1966), Vol. 2, pp. 5–21.

47. T. A. Germogenova and V. A. Utkin, in *Sbornik tezisov dokladov vsesoyuznoĭ konferentsii po zashite ot ioniziruyushikh izlucheniĭ yaderno-tekhnicheskikh ustanovok (Proceedings of the All-Union Scientific Conference on Ionizing-Radiation Shielding in Nuclear-Engineering Installations)*, 17–19 December 1974 (MIFI, Moscow, 1974), p. 74.

48. V. M. Kuvshinnikov and E. V. Pletnikov, in *Sbornik tezisov dokladov vsesoyuznoĭ nauchnoĭ konferentsii po zashite ot ioniziruyushikh izlucheniĭ yaderno-tekhnicheskikh ustanovok (Proceedings of the All-Union Scientific Conference on Ionizing-Radiation Shielding in Nuclear-Engineering Installations)*, 17–19 December 1974 (MIFI, Moscow, 1974), p. 77; in *Radiatsionnaya bezopasnost' i zashita AÉS (Radiation Safety and Shielding of Atomic Power Stations, No. 1)*, edited by Yu. A. Egorov *et al.* (Atomizdat, Moscow, 1976), pp. 46–50.

49. V. P. Mashkovich, in *Sbornik tezisov dokladov Vtoroĭ Vsesoyuznoĭ nauchnoĭ konferentsii po zashite ot ioniziruyushikh izlucheniĭ yaderno-tekhnicheskikh ustanovok (Proceedings of the Second All-Union Scientific Conference on Ionizing-Radiation Shielding in Nuclear-Engineering Installations)*, 19–21 December 1978 (MIFI, Moscow, 1978), p. 22; in *Radiatsionnaya bezopasnost' i zashita AÉS (Radiation Safety and Shielding of Atomic Power Stations, No. 5)*, edited by Yu. A. Egorov (Atomizdat, Moscow, 1981), pp. 187–199.

50. V. V. Bolyatko *et al.*, in *Sbornik tezisov dokladov Vtoroĭ Vsesoyuznoĭ nauchnoĭ konferentsii po zashite ot ioniziruyushikh izlucheniĭ yaderno-tekhnicheskikh ustanovok (Proceedings of the Second All-Union Scientific Conference on Ionizing-Radiation Shielding in Nuclear-Engineering Installations)*, 19–21 December 1978 (MIFI, Moscow, 1978), p. 43; in *Voprosy dozimetrii i zashity ot izlucheniĭ (Questions of Dosimetry and Radiation Shielding, No. 19)*, edited by V. I. Ivanov (Atomizdat, Moscow, 1980), pp. 109–129.

51. V. S. Avzyanov, V. P. Evdokimov, B. E. Lukhminskiĭ, and K. K. Yakushev, in *Sbornik tezisov dokladov Vtoroĭ Vsesoyuznoĭ nauchnoĭ konferentsii po zashite ot ioniziruyushikh izlucheniĭ yaderno-tekhnicheskikh ustanovok (Proceedings of the Second All-Union Scientific Conference on Ionizing-Radiation Shielding in Nuclear-Engineering Installations)*, 19–21 December 1978 (MIFI, Moscow, 1978), p. 76.

52. V. V. Bolyatko, M. Yu. Vyrskiĭ, A. A. Kamnev, and V. S. Troshin, in *Sbornik tezisov dokladov Vtoroĭ Vsesoyuznoĭ nauchnoĭ konferentsii po zashite ot ioniziruyushikh izlucheniĭ yaderno-tekhnicheskikh ustanovok (Proceedings of the Second All-Union Scientific Conference on Ionizing-Radiation Shielding in Nuclear-Engineering Installations)*, 19–21 December 1978 (MIFI, Moscow, 1978), p. 78.

53. T. A. Germogenova, V. G. Madeev, A. P. Suvorov, and V. A. Utkin, in *Sbornik tezisov dokladov Vtoroĭ Vsesoyuznoĭ nauchnoĭ konferentsii po zashite ot ioniziruyushikh izlucheniĭ yaderno-tekhnicheskikh ustanovok (Proceedings of*

the Second All-Union Scientific Conference on Ionizing-Radiation Shielding in Nuclear-Engineering Installations), 19–21 December 1978 (MIFI, Moscow, 1978), p. 80.

54. V. F. Khokhov, V. D. Tkachev, and I. N. Sheĭno, in Sbornik tezisov dokladov Vtoroĭ Vsesoyuznoĭ nauchnoĭ konferentsii po zashite ot ioniziruyushikh izlucheniĭ yaderno-technicheskikh ustanovok (Proceedings of the Second All-Union Scientific Conference on Ionizing-Radiation Shielding in Nuclear-Engineering Installations), 19–21 December 1978 (MIFI, Moscow, 1978), p. 90.

55. F. C. Maienschein, in Proceedings of the Fifth International Conference on Reactor Shielding, RSIC ORNL (Science Press, Princeton, NJ, 1977), pp. 1–10.

56. V. V. Bolyatko et al., in Sbornik tezisov dokladov Tret'eĭ vsesoyuznoĭ nauchnoĭ konferentsii po zashite ot ioniziruyushikh izlucheniĭ yaderno-tekhnicheskikh ustanovok (Proceedings of the Third All-Union Scientific Conference on Ionizing-Radiation Shielding in Nuclear-Engineering Installations), 27–29 October 1981 (TGU, Tbilisi, 1981), p. 65.

57. C. R. Weisbin, E. M. Oblow, F. R. Mynatt, and G. F. Flanagen, Transl. Am. Nucl. Soc. 22, 792 (1975).

58. GOST 15484-81. Izlucheniya ioniziruyushie i ikh izmereniya. Terminy i opredeleniya (All-Union State Standard 15484-81. Ionizing Radiations and Their Measurement. Terms and Definitions) (Izd. standartov, Moscow, 1981).

59. GOST 8.417-81. Edinitsy fizicheskikh velichin (All-Union State Standard 8.417-81. Units of Physical Quantities) (Izd. standartov, Moscow, 1981).

60. Metodicheskie ukazaniya "Vnedrenie i primenenie ST SEV 1052-78. Metrologiya. Edinitsy fizicheskikh velichin" (Methodological Instructions, "Introduction and Use of State Standard SEV 1052-78. Metrology. Units of Physical Quantities"), Report No. RD50-160-79 (Izd. standartov, Moscow, 1979).

61. Normy radiatsionnoĭ bezopasnosti NRB-76 (Radiation Safety Standards, NRB-76) (Atomizdat, Moscow, 1978).

62. V. I. Ivanov, V. P. Mashkovich, and É. M. Tsenter, Mezhdunarodnaya sistema edinits (SI) v atomnoĭ nauke i tekhnike. Spravochnoe rukovodstvo [The International System of Units (SI) in Atomic Science and Engineering. Handbook] (Energoizdat, Moscow, 1981).

63. N. G. Gusev, V. P. Mashkovich, and A. P. Suvorov, in Zashita ot ioniziruyushikh izlucheniĭ. Tom 1. Fizicheskie osnovy zashity (Shielding Against Ionizing Radiation. Vol. 1. The Physics of Shielding), 2nd ed., edited by N. G. Gusev (Atomizdat, Moscow, 1980).

64. V. I. Ivanov, Kurs dozimetrii (Course in Dosimetry), 3rd ed. (Atomizdat, Moscow, 1978).

65. Reactor Shielding for Nuclear Engineers, edited by N. M. Schaeffer (USAEC, Oak Ridge, TN, 1973).

66. Radiation quantities and units, ICRU Report No. 33, 1980.

67. Gosudarstvennava sistema obespecheniva edinstva izmereniĭ. Metrologiya. Terminy i opredeleniya. GOST 16263-70 (State System for Assuring the Unity of Measurements. Metrology. Terms and Definitions. All-Union State Standard 16263-70) (Gosudarstvennyĭ komitet standartov SM SSR, Moscow, 1972).

68. Gosudarstvennaya sistema obespecheniya edinstva izmereniĭ. Pokazateli tochnosti izmereniĭ i formy predstavleniya rezul'tatov izmereniĭ. GOST 8.011-72 (State System of Assuring the Unity of Measurements. Measures of the Accuracy of Measurements and Forms for Representing Experimental Results. All-Union State Standard 8.011-72) (Gosudarstvennyĭ komitet standartov SM SSSR, Moscow, 1972).

69. P. J. Kempion, D. E. Burns, and A. Williams, Practical Handbook on the Presentation of the Results of Measurements (Atomizdat, Moscow, 1979), Russian translation.

70. D. J. Hudson, *Lectures on Elementary Statistics and Probability* (CERN, Geneva, 1963), Russian translation published by Mir, Moscow, 1970.

71. L. Z. Rumshinskiĭ, *Matematicheskaya obrabotka rezul'tatov éksperimenta. Spravochnoe rukovodstvo (Mathematical Processing of Experimental Results. Handbook)* (Nauka, Moscow, 1971).

72. A. I. Abramov, Yu. A. Kazanskiĭ, and E. S. Matusevich, *Osnovy éksperimental'nykh metodov yadernoĭ fiziki (Fundamentals of Experimental Methods of Nuclear Physics)*, 2nd ed. (Atomizdat, Moscow, 1977).

73. V. I. Kalashnikova and M. S. Kozodaev, in *Detektory elementarnykh chastits (Elementary-Particle Detectors)*, edited by M. S. Kozodaev (Nauka, Moscow, 1966).

74. Yu. I. Balashov, V. V. Bolyatko, and A. I. Ilyushkin, in *Sbornik tezisov dokladov Tret'eĭ vsesoyuznoĭ nauchnoĭ konferentsii po zashite ot ioniziruyushikh izlucheniĭ yaderno-tekhnicheskikh ustanovok (Proceedings of the Third All-Union Scientific Conference on Ionizing-Radiation Shielding in Nuclear-Engineering Installations)*, 27–29 October 1981 (TGU, Tbilisi, 1981), p. 62.

75. T. Moorhead, in *Proceedings of an IAEA Seminar on the Physics of Fast and Intermediate Reactors, Vienna, 1961* (IAEA, Vienna, 1961), pp. 111–145.

76. A. Boioli and L. Fiorini, in *Proceedings of the Fifth International Conference on Reactor Shielding, RSIC ORNL* (Science Press, Princeton, NJ, 1977).

77. M. G. Kendall and A. Stuart, *Statistical Inference and Statistical Relationship* (Hafner, New York, 1968), Russian translation (Nauka, Moscow, 1973).

78. M. N. Nikolaev and B. G. Ryazanov, in *Voprosy atomnoĭ nauki i tekhniki. Seriya: Yadernye konstanty (Questions of Atomic Science and Engineering. Series on Nuclear Constants)* (TsNIIAI, Moscow, 1974), No. 17, pp. 21–40.

79. V. V. Nalimov, *Teoriya éksperimenta (Theory of Experiments)* (Nauka, Moscow, 1972).

80. A. A. Van'kov, A. I. Voropaev, and L. N. Yurova, *Analiz reaktorno-fizicheskogo éksperimenta (Analysis of Reactor-Physics Experiments)* (Atomizdat, Moscow, 1977).

81. E. M. Oblow, in *Proceedings of the Specialist's Meeting on Sensitivity Studies and Shielding Benchmarks, Paris, 7–10 October, 1975* (IAEA, Vienna, 1975).

82. M. Yu. Vyrskiĭ *et al.*, At. Energy **53**, 113 (1981).

83. V. Herrnberg, G. Hehn, and R. Nicks, in *Proceedings of the Specialist's Meeting on Sensitivity Studies and Shielding Benchmarks, Paris, 7–10 October 1975* (IAEA, Vienna, 1975), pp. 224–231.

84. H. A. Goldstein, in *Proceedings of the Fifth International Conference on Reactor Shielding, RSIC ORNL* (Science Press, Princeton, NJ, 1977), pp. 75–80.

85. M. Yu. Vyrskiĭ *et al.*, Preprint No. 124, Institute of Applied Mathematics, Academy of Sciences of the USSR, Moscow, 1976.

86. N. O. Bazazyants *et al.*, ARAMAKO-2F *sistema obespecheniya neĭtronnymi konstantami raschetov perenosa izlucheniya v reaktorakh i zashite (The ARAMAKO-2F System for Finding the Neutron Constants for Calculations on Radiation Transport in Reactors and Shielding. Manual)* (IPM AN SSR, Moscow, 1976).

87. M. Yu. Vyrskiĭ *et al.*, Preprint No. FEI-904, Obninsk, 1979.

88. A. A. Abagyan *et al.*, Preprint No. 122, Institute of Applied Mathematics, Academy of Sciences of the USSR, Moscow, 1978.

89. L. P. Abagyan, N. O. Bazazyants, I. I. Bondarenko, and M. N. Nikolaev, *Gruppovye konstanty dlya rascheta reaktorov (Group Constants for Reactor Calculations)* (Atomizdat, Moscow, 1964).

90. L. P. Abagyan *et al.*, *Rasprostranenie rezonansnykh neĭtronov v gomogennykh sredakh (Propagation of Resonance Neutrons in Homogeneous Media. Bulletin of the Information Center on Nuclear Data)* (Atomizdat, Moscow, 1968).

91. G. N. Manturov, in *Neĭtronnaya fizika. Materialy 5-ĭ Vsesoyuznoĭ konferentsii po neĭtronnoĭ fizike (Neutron Physics. Proceedings of the Fifth All-Union Conference on Neutron Physics, Kiev, 15–18 September 1980)*, edited by L. N. Usachev (TsNIIAI, Moscow, 1981), Pt. 3, p. 311.

92. L. P. Abagyan, N. O. Bazazyants, M. N. Nikolaev, and A. M. Tsibulya, in *Gruppovye konstanty dlya rascheta reaktorov i zashity. Spravochnik (Group Constants for Reactor and Shielding Calculations. Handbook)*, edited by M. N. Nikolaev (Énergoizdat, Moscow, 1981).

93. M. N. Nikolaev, in *Yadernye konstanty (Nuclear Constants), No. 8* (TsNIIAI, Moscow, 1972), p. 3.

94. V. V. Sinitsa and M. N. Nikolaev, Sistema SOKRATOR, podsistema GRUKON. Chast' 1. Printsipy organizatsii sistemy (The SOKRATOR System and the GRUKON Subsystem. Part 1. Organization Principles of the System), Report No. RD-16/050, 1978.

95. "Data formats and Procedures for the ENDR neutron cross-section library," edited by M. K. Drake, National Neutron Cross-Section Center, USAEC, Report No. BNL-50274, revised by D. Garber *et al.* (Report No. ENDF-102, 1976).

96. L. P. Abagyan, N. O. Bazazyants, M. N. Nikolaev, and A. M. Tsibulya, At. Energy **48**, 117 (1980).

97. V. V. Sinitsa *et al.*, in *Yaderno-fizicheskie issledovaniya v SSSR. Annotatsii programm (Nuclear-Physics Research in the USSR. Abstracts of Programs), No. 27* (Atomizdat, Moscow, 1979), p. 15.

98. T. A. Germogenova *et al.*, Preprint No. 140, IMP, Academy of Sciences of the USSR, 1979.

99. G. N. Manturov, Preprint No. FEI-1034, Obninsk, 1980.

100. T. A. Germogenova *et al.*, *Perenos bystrykh neĭtronov v ploskikh zashitakh (Fast-Neutron Transport in Plane Shields)* (Atomizdat, Moscow, 1971).

101. A. M. Voloshenko, E. I. Kostin, E. I. Panfilova, and V. A. Utkin, *ROZ-6 kompleks programm dlya resheniya uravneniya perenosa v odnomernykh geometriyakh (The ROZ-6 Program Library for Solving the Transport Equation in One-Dimensional Geometries. Manual)* (IPM AN SSSR, Moscow, 1980).

102. M. Yu. Vyrskiĭ and T. A. Germogenova, in *Voprosy dozimetrii i zashity ot izlucheniĭ (Questions of Radiation and Dosimetry Shielding)*, edited by V. K. Sakharov (Atomizdat, Moscow, 1976), pp. 82–89.

103. T. A. Germogenova *et al.*, in *Sbornik tezisov dokladov Tret'eĭ vsesoyuznoĭ nauchnoĭ konferentsii po zashite ot ioniziruyushikh izlucheniĭ yaderno-tekhnicheskikh ustanovok (Proceedings of the Third All-Union Scientific Conference on Ionizing-Radiation Shielding in Nuclear-Engineering Installations), 27–29 October 1981* (TGU, Tbilisi, 1981), p. 14.

104. A. I. Ilyushkin, in *Sbornik tezisov dokladov Vtoroĭ Vsesoyuznoĭ nauchnoĭ konferentsii po zashite ot ioniziruyushikh izlucheniĭ yaderno-tekhnicheskikh ustanovok (Proceedings of the Second All-Union Scientific Conference on Ionizing-Radiation Shielding in Nuclear-Engineering Installations), 19–21 December 1978* (MIFI, Moscow, 1978), p. 32.

105. T. A. Germogenova and A. I. Ilyushkin, in *Chislennoe reshenie uravneniya perenosa v odnomernykh zadachakh (Numerical Solution of the Transport Equation in One-Dimensional Problems)*, edited by T. A. Germogenova (IPM AN SSSR, Moscow, 1981), p. 116.

106. V. V. Korobeĭnikov, Preprint No. FEI-1039, Obninsk, 1980.

107. L. N. Alekseev, G. N. Manturov, and M. N. Nikolaev, At. Energy **49**, 221 (1980).

108. G. N. Manturov and M. N. Nikolaev, in *Neĭtronnaya fizika. Materialy 5-ĭ Vsesoyuznoĭ konferentsii po neĭtronnoĭ fizike (Neutron Physics. Proceedings of the Fifth All-Union Conference on Neutron Physics, Kiev, 15–19 September 1980)* (TsNIIAI, Moscow, 1981), Pt. 3, p. 316.

109. V. V. Bolyatko, A. I. Ilyushkin, V. P. Mashkovich, and V. K. Sakharov, in *Sbornik tezisov dokladov Tret'eĭ vsesoyuznoĭ nauchnoĭ konferentsii po zashite ot ioniziruyushikh izlucheniĭ yaderno-tekhnicheskikh ustanovok (Proceedings of the Third All-Union Scientific Conference on Ionizing-Radiation Shielding in Nuclear-Technology Installations), 27–29 October 1981* (TGU, Tbilisi, 1981), p. 66.

110. V. V. Bolyatko *et al.*, in *Sbornik tezisov dokladov Tret'eĭ vsesoyuznoĭ nauchnoĭ konferentsii po zashite ot ioniziruyushikh izlucheniĭ yaderno-tekhnicheskikh ustanovok (Proceedings of the Third All-Union Scientific Conference on Ionizing-Radiation Shielding in Nuclear-Technology Installations), 27–29 October 1981* (TGU, Tbilisi, 1981), p. 64.

111. Y. Huang and H. Goldstein, Transl. Am. Nucl. Soc. **18**, 365 (1974).

112. A. K. McCracken and M. G. Grimstone, in *Proceedings of the Specialist's Meeting on Sensitivity Studies and Shielding Benchmarks, Paris, 7–10 October 1975* (IAEA, Vienna, 1975), pp. 175–208.

113. G. Hehn *et al.*, in *Proceedings of the Specialist's Meeting on Sensitivity Studies and Shielding Benchmarks, Paris, 7–10 October 1975* (IAEA, Vienna, 1975), pp. 130–147.

114. H. Werle *et al.*, in *Proceedings of the Specialist's Meeting on Sensitivity Studies and Shielding Benchmarks, Paris, 7–10 October 1975* (IAEA, Vienna, 1975), pp. 98–103.

115. V. V. Bolyatko *et al.*, in *Sbornik tezisov dokladov Tret'eĭ vsesoyuznoĭ nauchnoĭ konferentsii po zashite ot ioniziruyushikh islucheniĭ yaderno-tekhnicheskikh ustanovok (Proceedings of the Third All-Union Scientific Conference on Ionizing-Radiation Shielding in Nuclear-Technology Installations), 27–29 October 1981* (TGU, Tbilisi, 1981), p. 64.

116. D. E. Bartine *et al.*, *ORNL cross-section sensitivity analysis applications for radiation shielding*, Los Alamos Scientific Laboratory, Reports No. CONF-740903-8 and No. LA-UR-74-600.

117. D. E. Bartine, E. M. Oblow, and F. R. Mynatt, in Proceedings of a National Topical Meeting on New Developments in Reactor Physics and Shielding, Kiamesha Lake, NY, 1972 (unpublished), p. 512.

118. E. M. Oblow and R. M. Maerker, paper to be presented at FBR Shielding Seminar at Obninsk, USSR, 13–17 November 1978.

119. V. V. Bolyatko, M. Yu. Byrskiĭ, and A. A. Stroganov, in Radiatsionnaya bezopasnost' AÉS. Trudy VTI (Radiation Safety at Atomic Power Plants. Proceedings of the VTI) No. 26, Moscow, 1979, pp. 22–28.

120. S. D. Lympany, A. K. McCracken, and A. Packwood, in *Proceedings of the Specialist's Meeting on Differential and Integral Nuclear Data Requirements for Shielding Calculations, Vienna, 12–15 October 1976* (IAEA, Vienna, 1978), pp. 108–111, 175–195.

121. A. F. Avery and S. D. Lympany, in *Proceedings of the Specialist's Meeting on Sensitivity Studies and Shielding Benchmarks, Paris, 7–10 October 1975* (IAEA, Vienna, 1975), pp. 248–263.

122. G. Brandicourt and C. Devillers, in *Proceedings of the Specialist's Meeting on Differential and Integral Nuclear Data Requirements for Shielding Calculations, Vienna, 12–15 October 1976* (IAEA, Vienna, 1978), pp. 91–124.

123. G. Minsart and C. Van Bosstraeten, in *Proceedings of the Specialist's Meeting on Differential and Integral Nuclear Data Requirements for Shielding Calculations, Vienna, 12–15 October 1976* (IAEA, Vienna, 1978), pp. 155–174.

124. U. Canali, G. Gonano, and P. Nicks, in *Proceedings of the Specialist's Meeting on Differential and Integral Nuclear Data Requirements for Shielding Calculations, Vienna, 12–15 October 1976* (IAEA, Vienna, 1978), pp. 125–145.

125. G. Hehn and J. Rataj, in *Proceedings of the Specialist's Meeting on Differential and Integral Nuclear Data Requirements for Shielding Calculations, Vienna, 12–15 October 1976* (IAEA, Vienna, 1978), pp. 275–283.

126. U. Canali, G. Gonano, and R. Nicks, in *Proceedings of the Specialist's Meeting on Differential and Integral Nuclear Data Requirements for Shielding Calculations, Vienna, 12–15 October 1976* (IAEA, Vienna, 1978), pp. 147–154.

127. S. D. Lympany, A. K. McCracken, and A. Packwood, in *Proceedings of the Specialist's Meeting on Differential and Integral Nuclear Data Requirements for Shielding Calculations, Vienna, 12–15 October 1976* (IAEA, Vienna, 1978), pp. 285–322.

128. V. V. Bolyatko *et al.*, in *Radiatsionnaya zashita na atomnykh élektrostantsiyakh (Radiation Shielding at Atomic Power Plants)*, edited by A. P. Suvorov and S. G. Ts'shin (Atomizdat, Moscow, 1978).

129. T. A. Germogenova, V. I. Zhuravlev, A. P. Suvorov, and V. A. Utkin, in *Voprosy fiziki zashity reaktorov (Questions of the Physics of Radiator Shielding)*, edited by D. L. Broder *et al.* (Atomizdat, Moscow, 1972), No. 5, pp. 22–46.

130. F. R. Mynatt, in *Handbook on Radiation Shielding for Engineers*, Russian translation edited by D. L. Broder *et al.* (Atomizdat, Moscow, 1972).

131. L. D. Lathrop, DTF-IV, *A Fortran-IV Program for Solving the Multigroup Transport Equation with Anisotropic Scattering*, Los Alamos Scientific Laboratory Report No. LA-3373, 1965.

132. R. Vaidyanathan, Nucl. Sci. Eng. **76**, 46 (1979).

133. P. Barbucci and F. Di Pavguantonio, Nucl. Sci. Eng. **63**, 179 (1977).

134. D. V. Gopinath, A. Natarajan, and V. Sundararaman, Nucl. Sci. Eng. **75**, 181 (1980).

135. E. S. Lee and R. Vaidynathan, Nucl. Sci. Eng. **76**, 1 (1979).

136. E. W. Larsen and W. G. Miller, Nucl. Sci. Eng. **73**, 76 (1980).

137. R. E. Alcouffe, E. M. Larsen, W. F. Miller, and B. R. Wienke, Nucl. Sci. Eng. **71**, 111 (1979).

138. W. R. Martin and J. J. Duderstadt, Nucl. Sci. Eng. **62**, 371 (1977).

139. K. Takeuchi, J. Nucl. Sci. Technol. **8**, 141 (1971).

140. N. Sasamoto and I. Takeuchi, Nucl. Sci. Eng. **17**, 330 (1979).

141. T. A. Germogenova, in *Voprosy teorii perenosa izlucheniya (Questions of Radiation Transport Theory)* (IPM AN SSSR, 1981), p. 5.

142. A. M. Voloshenko and T. A. Germogenova, in *Voprosy teorii perenosa izluchenii (Questions of Radiation Transport Theory)* (IPM AN SSSR, 1981).

143. A. M. Vodoshenko, in *Voprosy teorii perenosa izluchenii (Questions of Radiation Transport Theory)* (IPM AN SSSR, 1981).

144. A. P. Suvorov and V. A. Utkin, in *Yadernye konstanty (Nuclear Constants)*, edited by V. A. Kuznetsov *et al.* (Atomizdat, Moscow, 1971), No. 7, pp. 394–407.

145. F. Jauho and H. Kalli, Nucl. Sci. Eng. **31**, 318 (1968).

146. J. A. Jung, Nucl. Sci. Eng. **65**, 130 (1978).

147. W. J. Mikols and J. K. Shultis, Nucl. Sci. Eng. **63**, 91 (1977).

148. K. D. Lathrop, Nucl. Sci. Eng. **32**, 357 (1968).

149. W. F. Miller and W. H. Reed, Nucl. Sci. Eng. **63**, 391 (1977).

150. L. P. Bass, Preprint No. 14, Institute of Applied Mathematics, Academy of Sciences of the USSR, 1975.

151. K. Shure, Nucl. Sci. Eng. **19**, 310 (1954).

152. L. P. Bass, T. A. Germogenova, and A. P. Suvorov, in *Voprosy fiziki zashity reaktorov (Questions of the Physics of Reactor Shielding)*, edited by L. D. Broder *et al.* (Atomizdat, Moscow, 1969), No. 4, p. 203.

153. J. J. Sapyta and G. L. Simmons, Nucl. Technol. **26**, 508 (1975).

154. L. Hjarue and M. Leimdorfer, Nucl. Sci. Eng. **24**, 165 (1966).

155. N. M. Korobov, Dokl. Akad. Nauk SSSR **124**, 1207 (1959).

156. A. P. Veselkin *et al.*, *Inzhenernyĭ raschet zashity atomnykh élektrostantsiĭ (Engineering Calculations on the Shielding at Atomic Power Plants)* (Atomizdat, Moscow, 1976).

157. J. P. Odom and J. K. Shultis, Nucl. Sci. Eng. **59**, 278 (1976).

158. W. J. Mikols and J. K. Shultis, Nucl. Sci. Eng. **62**, 739 (1977).

159. M. Yu. Vyrskiĭ *et al.*, in *Radiatsionnaya bezopasnost' i zashita AÉS (Radiation Safety and Shielding at Atomic Power Plants)*, edited by Yu. A. Egorov *et al.* (Atomizdat, Moscow, 1980), No. 4, pp. 209–219.

160. V. F. Khokhlov, V. D. Tkachev, V. L. Reĭtblat, and I. N. Sheĭno, At. Energy **44**, 324 (1978).

161. M. M. Savos'kin and M. Yu. Vyrskiĭ, in *Voprosy atomnoĭ nauki i tekhniki. Ser. Yadernye konstanty (Questions of Atomic Science and Engineering. Series on Nuclear Constants)* (TsNIIAI, Moscow, 1980), No. 2(37) pp. 107–111.

162. M. N. Zizin, *Raschet neĭtronno-fizicheskikh kharakteristik reaktorov na bystrykh neĭtronakh (Calculation of the Neutron-Physics Characteristics of Fast-Neutron Reactors)* (Atomizdat, Moscow, 1978).

163. L. A. Trykov, V. P. Semenov, and A. N. Nikolaev, At. Energy **39**, 56 (1975).

164. V. D. Tkachev and V. F. Khokhlov, in *Radiatsionnaya bezopasnost' i zashita AÉS (Radiation Safety and Shielding at Atomic Power Plants)* (Atomizdat, Moscow, 1976), No. 2, pp. 197–199.

165. M. Yu. Vyrskiĭ, I. I. Linge, and A. I. Ilyushkin, *Organizatsiya raschetnykh issledovaniĭ odnomernykh zashit po programme ROZ-9 (Organization of Computational Studies of One-Dimensional Shielding by the ROZ-9 Program)* (MIFI, Moscow, 1980); Deposit No. 5423-80, manuscript deposited with the All-Union Institute of Scientific and Technological Information, 22 December 1980.

166. E. Oblow, J. Ching, and J. Drischer, Transl. Am. Nucl. Soc. **17**, 547 (1973).

167. V. A. Herrnberger, in *Proceedings of the Fifth International Conference on Reactor Shielding, RSIC ORNL* (Science Press, Princeton, NJ, 1977).

168. M. Yu. Vyrskiĭ *et al.*, in *Sbornik tezisov dokladov Tret'eĭ vsesoyuznoĭ nauchnoĭ konferentsii po zashite ot ioniziruyushikh izlucheniĭ yaderno-tekhnicheskikh ustanovok (Proceedings of the Third All-Union Scientific Conference on Ionizing-Radiation Shielding in Nuclear-Engineering Installations)*, 27–29 October 1981 (TGU, Tbilisi, 1981), p. 52.

169. M. Yu. Vyrskiĭ, T. A. Germogenova, A. I. Ilyushkin, and I. I. Linge, in *Sbornik tezisov dokladov Vtoroĭ Vsesoyuznoĭ nauchnoĭ konferentsii po zashite ot ioniziruyushikh izlucheniĭ yaderno-tekhnicheskikh ustanovok (Proceedings of the Second All-Union Scientific Conference on Ionizing-Radiation Shielding in Nuclear-Engineering Installations)*, 19–21 December 1978 (MIFI, Moscow, 1978), p. 10.

170. M. Yu. Vyrskiĭ *et al.*, in *Sbornik tezisov dokladov Tret'eĭ vsesoyuznoĭ nauchnoĭ konferentsii po zashite ot ioniziruyushikh izlucheniĭ yaderno-tekhnicheskikh ustanovok (Proceedings of the Third All-Union Scientific Conference on Ionizing-Radiation Shielding in Nuclear-Engineering Installations)*, 27–29 October 1981 (TGU, Tbilisi, 1981), p. 53.

171. J. Butler, in *Proceedings of the Fifth International Conference on Reactor Shielding, RSIC ORNL* (Science Press, Princeton, NJ, 1977), pp. 120–131.

172. J. C. Estiot and J. P. Trapp, in *Proceedings of the Specialist's Meeting on Differential and Integral Nuclear Data Requirements for Shielding Calculations, Vienna, 12–15 October 1976* (IAEA, Vienna, 1978), pp. 221–241.

173. E. Greenspan, Y. Karri, and D. Gilai, in *Proceedings of a Seminar-Workshop of Theory and Application of Sensitivity and Uncertainty Analysis, 22–24 August 1978, Oak Ridge, Tennessee,* ORNL Report No. ORNL/RSIC-24, pp. 231–250.

174. J. Butler, paper presented at the Meeting of the European-American Committee on Reactor Physics, Richland, WA, July 1970.

175. V. P. Mashkovich and S. G. Tsypin, At. Energy **38**, 398 (1975).

176. J. Butler, ESIS Newsletter (No. 18), July 1976.

177. J. Butler and A. K. McCracken, *The NEA/EURATOM Restricted Meeting on Specialists on Shielding Benchmark Experiments Held at AEE Winfrith, April, 1975,* Report No. NEACRP-L-121.

178. *Proceedings of the Joint IAEA-OECD Technical Committee Meeting on Differential and Integral Nuclear Data Requirements for Shielding Calculations, Vienna, Austria, 12–15 October 1976,* Report No. IAEA-207 (IAEA, Vienna, 1978).

179. R. Nicks and J. Butler, Summary Report in *Proceedings of the Specialist's Meeting on Sensitivity Studies and Shielding Benchmarks, Paris, 7–10 October 1975* (IAEA, Vienna, 1975), pp. 6–13.

180. R. Nicks, in *Proceedings of the Specialist's Meeting on Sensitivity Studies and Shielding Benchmarks, Paris, 7–10 October 1975* (IAEA, Vienna, 1975), pp. 19–22.

181. M. Martini and L. Bozzi, ESIS Newsletter, Special Issue No. 2, Ispra, March 1974.

182. R. E. Maerker and F. J. Muckenthaler, Nucl. Sci. Eng. **52**, 227 (1973).

183. R. E. Maerker and F. J. Muckenthaler, *Final Report on a Benchmark Experiment for Neutron Transport through Iron and Stainless Steel,* ORNL Report No. ORNL-4892, 1974.

184. R. E. Maerker *et al., Final Report on a Benchmark Experiment for Neutron Transport in Thick Sodium,* ORNL Report No. ORNL-4880, 1974.

185. Oka Yoshiaki *et al., Report on Joint NEA/IAEA Specialists Meeting on Differential and Integral Nuclear Data Requirements for Shielding Calculations, Vienna, October 1976* (IAEA, Vienna, 1976).

186. S. An *et al.,* in *Proceedings of the Specialist's Meeting on Sensitivity Studies and Shielding Benchmarks, Paris, 7–10 October 1975* (IAEA, Vienna, 1975), pp. 104–110.

187. M. D. Carter and A. Packwood, in *Proceedings of the Specialist's Meeting on Sensitivity Studies and Shielding Benchmarks, Paris, 7–10 October 1975* (IAEA, Vienna, 1975), pp. 111–119.

188. Y. Furuta, N. Sasamoto, and S. Tanaka, in *Proceedings of the Specialist's Meeting on Sensitivity Studies and Shielding Benchmarks, Paris, 7–10 October 1975* (IAEA, Vienna, 1975), pp. 120–126.

189. L. A. Trykov *et al.,* in *Voprosy dozimetrii i zashity ot izluchenii (Questions of Radiation and Dosimetry Shielding)* (Atomizdat, Moscow, 1979), No. 18, pp. 93–98.

190. L. A. Trykov *et al.,* Preprint No. FEI-943, Obninsk, 1979.

191. L. A. Trykov *et al.,* Preprint No. FEI-1096, Obninsk, 1980.

192. R. Nicks and G. Perlini, ESIS Newsletter, Special Issue No. 1, Ispra, March 1973.

193. F. C. Maienschein, R. E. Maerker, and F. J. Muckenthaler, FBR Shielding Seminar, Obninsk, USSR, 13–17 November 1978.

194. V. V. Bolyatko *et al.*, in *Radiatsionnaya bezopasnost' i zashita AÉS (Radiation Safety and Shielding at Atomic Power Plants)* (Atomizdat, Moscow, 1980), No. 4, pp. 70–78.

195. J. Y. Barre, "Fast reactors: Definition of a Standard Configuration for the Comparison of Shielding Calculations," in NEA Meeting on Sensitivity Studies and Shielding Benchmarks at Paris, October 1975 (note circulated to participants).

196. G. Hehn and J. Koban, ESIS Newsletter, Special Issue No. 4, Ispra, January 1976.

197. J. Bishop, K. Smitton, and A. Packwood, in *Questions of Reactor Shielding. No. 3. Proceedings of the Fourth International Conference on Reactor Shielding, Paris, October 1972*, Russian translation (Atomizdat, Moscow, 1974), pp. 25–38.

198. L. F. Hansen, J. D. Anderson, P. S. Brown, R. J. Howerton, J. L. Kammerdicher, C. M. Logan, E. F. Plechaty, and C. Wong, Nucl. Sci. Eng. **51**, 278 (1973).

199. V. S. Troshin and E. A. Kramer-Ageev, At. Energy No. 1, 37 (1970).

200. E. A. Kramer-Ageev, V. S. Troshin, and E. G. Tikhonov, *Aktivatsionnye metody spektrometrii neĭtronov (Activation Methods of Neutron Spectrometry)* (Atomizdat, Moscow, 1976).

201. E. M. Oblow and C. R. Weisbin, in *Proceedings of the Fifth International Conference on Reactor Shielding* (Science Press, Princeton, NJ, 1977), pp. 140–148.

202. W. A. Rhoades and F. R. Mynatt, *The DOT-III Two-Dimensional Discrete Ordinates Transport Code*, ORNL Report No. ORNL-TM-4280, 1973.

203. C. R. Weisbin *et al.*, Application of FORSS Sensitivity and Uncertainty Methodology to Fast Reactor Benchmark Analysis, ORNL Report No. ORNL/TM-5563, 1976.

Printed in the United States
By Bookmasters